Human Resources in the Foodservice Industry: Organizational Behavior Management Approaches

Human Resources in the Foodservice Industry: Organizational Behavior Management Approaches has been co-published simultaneously as *Journal of Foodservice Business Research*, Volume 9, Numbers 2/3 2006.

Monographic Separates from the *Journal of Foodservice Business Research™*

For additional information on these and other Haworth Press titles, including descriptions, tables of contents, reviews, and prices, use the QuickSearch catalog at http://www.HaworthPress.com.

The *Journal of Foodserice Business Research™* is the successor title *Journal of Restaurant & Foodservice Marketing™*,* which changed title after Vol. 4, No. 4, 2001. The *Journal of Foodserice Business Research™*, under its new title, begins with Vol. 5, No, 1, 2002.

Human Resources in the Foodservice Industy: Organizational Behavior Management Approaches, edited by Dennis Reynolds, PhD, and Karthik Namasivayam, PhD (Vol 9, No. 2/3, 2006). *Examination of the latest empirical research on organizational behavior and human resources management in the foodservice industry.*

****Quick Service Restaurants, Franchising, and Multi-Unit Chain Management,*** edited by H. G. Parsa, PhD, FMP, and Francis A. Kwansa, PhD (Vol. 4, No. 3/4, 2001). *"Fills an important void, bringing together a much-needed unique collection of the latest research. This is important reading for experienced management personnel within the Quick Service Restaurant industry as well as for graduate students and upper-level undergraduates." (Audrey C. McCool, EdD, RD, FADA, FMP, Assistant Dean for Research, William F. Harrah College of Hotel Administration, University of Nevada, Las Vegas)*

Human Resources in the Foodservice Industry: Organizational Behavior Management Approaches

Dennis Reynolds, PhD
Karthik Namasivayam, PhD
Editors

Human Resources in the Foodservice Industry: Organizational Behavior Management Approaches has been co-published simultaneously as *Journal of Foodservice Business Research*, Volume 9, Numbers 2/3 2006.

CRC Press is an imprint of the
Taylor & Francis Group, an **informa** business

CRC Press
Taylor & Francis Group
6000 Broken Sound Parkway NW, Suite 300
Boca Raton, FL 33487-2742

Visit the Taylor & Francis Web site at
http://www.taylorandfrancis.com

and the CRC Press Web site at
http://www.crcpress.com

Human Resources in the Foodservice Industry: Organizational Behavior Management Approaches

CONTENTS

ABOUT THE EDITORS

Dennis Reynolds is Associate Director and the Ivar Haglund Distinguished Professor of Hospitality Management at the Washington State University School of Hospitality Business Management, teaches and conducts research on management topics of interest to the hospitality industry. His teaching–targeting undergraduate, graduate, and executive-education students–centers on global service-management issues.

Dr. Reynolds is a frequent speaker to management groups in Asia, Europe, and North America and has been cited in various media (including *Morning Edition* on National Public Radio) as well as newspapers and magazines around the globe. His lively seminars cover such topics as maximizing productivity in the workplace and understanding the key value drivers in the hospitality industry. Recent talks have also addressed complex linkages between service and profit in a variety of settings and industries including sport management, education, and healthcare.

Professor Reynolds's research focuses on pathways leading to enhanced managerial efficiency and effectiveness, especially in service organizations, through the application of operations-management tools and techniques. Recent papers have also addressed the related effects of management feedback on subordinate self-efficacy and a new approach to evaluating operational efficiency for multiunit restaurant organizations. His work, which has captured numerous Academy of Management and other awards, has been published in such journals as the *Journal of Foodservice Business Research,* the *Advanced Management Journal*, the *Cornell Hotel and Restaurant Administration Quarterly*, the *Journal of Hospitality and Tourism Research*, and the *International Journal of Hospitality Management*. He is the author of *On-Site Foodservice Management: A Best Practices Approach* (Wiley & Sons, 2003). He is also the creator of a restaurant simulation titled *Restaurateur*®; this 'game' utilizes the allure of restaurant ownership to foster students' entrepreneurial interests by demonstrating how a restaurant concept can become a viable business venture.

Prior to joining WSU's College of Business, Dr. Reynolds was the J. Thomas Clark Professor of Entrepreneurship and Personal Enterprise at the Cornell University School of Hotel Administration, at which he received numerous honors, including the 2002 Teacher of the Year Award. He holds a doctoral degree from Cornell University in hospitality management and a Master of Professional Studies degree with a concentration in management operations also from Cornell, as well as a Bachelor of Science degree in hotel, restaurant, and institutional management from Golden Gate University.

Karthik Namasivayam is an Assistant Professor at the School of Hospitality Management, The Pennsylvania State University where he teaches Human Resources Management and Organizational Behavior. He also teaches Leadership and Change Management at the graduate and undergraduate levels. His research explores the intersections between the hospitality organization and the consumer.

Dr. Namasivayam explores the notion of consumer satisfaction in his research. The main thrust of his research is in connecting organizational human resource practices and organizational structures to consumer behavior. He has also, more recently, explored the notion of innovation and knowledge management in hospitality organizations. His publications have been widely recognized, with the most recent recognition being the W. Bradford Wiley Best Paper of the Year Award for 2005 awarded by the I-CHRIE. He is widely published including *Psychology & Marketing*, the *International Journal of Service Industries Management*, the *Cornell Hotel and Restaurant Administration Quarterly*, the *Journal of Human Resources in Hospitality and Tourism*, *Journal of Travel Research*, the *International Journal of Hospitality Management*, the *Journal of Intellectual Capital*, the *Journal of Hospitality and Tourism Research*, and the *Journal of Foodservice Business Research*.

Dr. Namasivayam draws on his over 25 years of worldwide hospitality industry experience–as an entrepreneur, manager, frontline worker, and project consultant in three continents: Asia, Europe and America–to enrich both his teaching and his research. Dr. Namasivayam holds a doctoral degree in hospitality management from Cornell University and a Master's of Management in Hospitality also from Cornell University. He obtained his Bachelor's degree in Economics from Madras University, India.

Acknowledgments

The editors thank Dr. H.G. Parsa, Honorary Editor-In-Chief for his support and guidance in putting together this Special Double Issue. The support of Dr. David Cranage, Editor, is also gratefully acknowledged. This issue would not have been possible without the help of a dedicated group of reviewers. The editors thank the following individuals for their diligence and patience through the process.

Dr. Albert (Bart) Bartlett, *The Pennsylvania State University*
Dr. Mark Beattie, *Washington State University*
Dr. Shirley Gilmore, *Iowa State University*
Dr. Dogan Gursoy, *Washington State University*
Dr. Bharath M. Josiam, *University of North Texas*
Dr. Darin Leeman, *UBS*
Dr. Edward Merritt, *California State University Pomona*
Dr. Michael O'Fallon, *James Madison University*
Dr. Swathi Ravichandran, *Kent State University*
Dr. Catherine Strohbehn, *Iowa State University*
Dr. Xinyuan Zhao, *South China University of Technology*

Dennis Reynolds, Washington State University
Karthik Namasivayam, The Pennsylvania State University
Guest Editors

Organizational Behavior and Human Resource Management in the Global Foodservice Industry: An Introduction

Dennis Reynolds
Karthik Namasivayam

We find ourselves in an era marked by sweeping political, economic, social, and technological change. It would be easy in such an environment to overlook another sweeping change, a seemingly mundane transformation in the fundamental core of human experience: our eating habits. With increasing prosperity, human beings are embracing the convenience, epicurean pleasure, and value of eating meals outside of the home. In absolute terms, more money is spent outside the home on foodservice than at any other time in history, but the far more telling statistic is the steady rise in the *percentage of our food dollars* that we spend at restaurants.

In 1955 individuals spent, on average, just 25% of their food dollars outside of the home in the United States, but the National Restaurant

---

Dennis Reynolds, PhD, is Associate Director and the Ivar B. Haglund Endowed Chair in Hospitality Business Management, School of Hospitality Business Management, Washington State University, Todd Add 477, Pullman, WA 99164-4742 (E-mail: der@wsu.edu).

Karthik Namasivayam, PhD, is Assistant Professor, School of Hospitality Management, The Pennsylvania State University, 213 Mateer Building, University Park, PA 16802 (E-mail: kun1@psu.edu).

[Haworth co-indexing entry note]: "Organizational Behavior and Human Resource Management in the Global Foodservice Industry: An Introduction." Reynolds, Dennis, and Karthik Namasivayam. Co-published simultaneously in *Journal of Foodservice Business Research* (The Haworth Hospitality & Tourism Press, an imprint of The Haworth Press, Inc.) Vol. 9, No. 2/3, 2006, pp. 1-5; and: *Human Resources in the Foodservice Industry: Organizational Behavior Management Approaches* (ed: Dennis Reynolds, and Karthik Namasivayam) The Haworth Hospitality & Tourism Press, an imprint of The Haworth Press, 2006, pp. 1-5. Single or multiple copies of this article are available for a fee from The Haworth Document Delivery Service [1-800-HAWORTH, 9:00 a.m. - 5:00 p.m. (EST). E-mail address: docdelivery@haworthpress.com].

Available online at http://jfbr.haworthpress.com

doi:10.1300/J369v09n02_01

Association now predicts that, during the current decade, this number will exceed 50% (NRA, 2006). Worldwide, annual consumer expenditures on restaurant visits during the same time frame should exceed one trillion dollars.

These changing patterns in consumer behavior and expenditures affect the foodservice business in many ways, the most fundamental of which is the manner in which resources are managed. In order of importance and complexity, in a business in which service is paramount, human resources naturally attract the most scholarly attention. It is no surprise that the manner in which people are managed has a causal effect on the profitability of today's restaurant venture (Harris & Bonn, 2000).

Understanding both individual behavior and group dynamics–organizational behavior–in restaurants is therefore critical for researchers and practitioners. As Podsakoff and MacKenzie (1997) have argued, knowing how people interact and function in the workplace allows operators to fully leverage their human capital. This knowledge also allows firms to differentiate themselves on the basis of service, the fundamental differentiator in all segments of the service industry.

For this reason, research in organizational behavior and human resource management yields substantive results for the foodservice industry. Scholars performing strong conceptual and empirical work in these areas are creating a stronger theoretical understanding of the various linkages involved. Such research also spawns novel managerial applications and approaches. For foodservice managers–given the critical importance of human capital in a competitive global marketplace–such research will produce managerial solutions that increase efficiency, employee and consumer satisfaction, and–ultimately–organizational success.

This, then, is the impetus driving the collection of work offered here. Researchers representing several countries and diverse backgrounds have contributed well supported empirical results, unique insights, and ontological deductions. Beginning with a macro-scale perspective on the global foodservice industry, the articles eventually address more micro-oriented topics, each with unique relevance to organizational behavior and human resource management in restaurants.

Chris Muller and Robin DiPietro take the macro perspective, focusing particularly on chain restaurant operations as they introduce a theoretical framework for multi-unit management development. The researchers lay a foundation for future study by proposing a model that integrates a series of phases in the development of success characteristics. This research is particularly important given the growing need for multi-unit managers around the world.

Next, Bonnie Canziani reviews the literature in the fields of business and law in order to determine the legal, business, and ethical issues at stake in setting language policies for personnel in foodservice businesses within the United States. The result is a best practices approach to managing a linguistically diverse workforce. Although the U.S. legal system forms the backdrop, the results apply to foodservice operations anywhere in the world.

Continuing with the effects created by globalization, Van Dyk et al. takes a different approach to the managerial implications associated with managing a diverse workforce, one that is often comprised of non-native speakers doing food preparation work. In particular, the study examines the relationship between employee self-efficacy in English fluency and employee perceptions of organizational support and organizational commitment. These constructs are vital to effective foodservice management, as demonstrated by previous studies that have provided evidence confirming that perceived organizational support and organizational commitment are important antecedents of several organizational and employee outcomes including job satisfaction, performance, and turnover intentions. The takeaway from the study is that language barriers in the workplace require special attention to ensure that employees feel connected to a foodservice organization.

Focusing on the ever-challenging logistics of serving hot food hot and cold food cold, while acknowledging the difficulties associated with the foodservice industry's unique workforce composition, Rhiannon Fante, Leslie Shier, and John Austin report the results of a unique field experiment. The researchers evaluated the effects of task clarification and self-monitoring on low-frequency food preparation behaviors of kitchen employees. The results suggest that task clarification alone–that is, simply providing information about a specific task–is not sufficient to effectively increase performance; conversely, self-monitoring is a viable approach to increasing the frequency of such food preparation behaviors as conducting food temperature checks.

The first half of the text concludes with a novel look at restaurant sales contests, a popular incentive program used by foodservice operators to increase check average and total revenue. David Corsun, Amy McManus, and Clark Kincaid note that such sales contests are adopted in a variety of foodservice settings, but they question the efficacy of this purportedly sales-boosting approach. In this conceptual paper, the authors provide intriguing cause for concern by examining the basic tenets that underlie the related applications.

The second half of this work maintains our global perspective but shifts the focus to more specific issues in organizational behavior and human resource management. For example, Jane Barnes and Susan Fisher consider the importance of collaboration in the new economy, addressing the barriers that cultural differences may create in any collaborative venture. Using nonprofit organizations as a backdrop, the authors adopt a case-study approach in examining the role that culture may play in forming cross-cultural business alliances.

Next, Yongmei Liu and Jixia Yang present a conceptual model based on the proposition that foodservers' emotional experiences and affective service delivery are mediating mechanisms through which the climate for service leads to favorable customer outcomes. The authors propose four mechanisms: motivation, capability, carryover, and compensation. They argue persuasively that a strong climate for service enhances the emotional quality of foodservers' experiences at work. Such positive emotional experiences encourage employees to develop better affective service delivery skills; these positive emotions may also be carried over to the service encounter. The authors suggest recommendations that managers can use to create a positive work climate and identify related research topics.

Adopting an empirical approach to a related issue, Hsu-I Huang embarked on a study of restaurant cooks in Taiwan to learn how they differ on the basis of locus of control, job satisfaction, work stress, and turnover intentions, depending on each participant's demographic profile. In this endeavor, Huang has identified a typology involving three 'types' of cooks, each with different attitudes associated with the aforementioned constructs. The paper then builds on this finding with suggestions for developing effective human-resource strategies, particularly for back-of-the-house employees.

Robin DiPietro and Merwyn Strate also consider foodservice workers' attitudes, but focus on the perceptions of quick-service-restaurant managers regarding older workers. Citing the need to broaden the pool of candidates for line positions in the segment, the authors present a case for considering older workers as a viable labor source. Interestingly, they found that their small sample of respondents supported the concept, but that the associated attitudes did not produce the expected corresponding behaviors.

Finally, Vinnie Jauhari and Kamal Manaktola also consider the importance of finding qualified employees but focus their study on student internships. They explain how internships serve as appropriate job previews for students and can be used to create positive employment experiences, possibly even creating loyalty in virtue of which the intern becomes a future customer. The paper's most important contribution, however, is its comparison of interns' experiences in the United Kingdom and India. These were compared on the basis of training, mentorship, hygiene factors, rewards, and recognition. The result is a useful model for managing internship experiences within the foodservice industry.

The thread that ties this body of research together, the criticality of effective restaurant management, is underscored by much-publicized failure rates in an environment marked by rapid, constant change. To increase organizational effectiveness and bring new understanding of the various linkages to the fore, we must therefore continue, as Davis (1971) so artfully argued, to explore theories that deny certain assumptions or contradict commonly accepted practices. Our inclusion of such inquiry here will help create a global foodservice industry of the future that reflects its humble beginnings but heralds the most innovative–and provocative–organizational behavior and human resource management approaches.

REFERENCES

Davis, M. S. (1971). That's interesting! Towards a phenomenology of sociology and sociology of phenomenology. *Philosophy of the Social Sciences, 1*, 309-344.

Harris, K. J., & Bonn, M. A. (2000). Training techniques and tools: Evidence from the foodservice industry. *Journal of Hospitality & Tourism Research, 24*, 320-335.

National Restaruant Association (2005). 2006 Restaurant industry forecast. Washington, D.C.: NRA.

Podsakoff, P. M., & MacKenzie, S. B. (1997). Impact of organizational citizenship behavior on organizational performance: A review and suggestion for future research. *Human Performance, 10*(2), 133-151.

doi:10.1300/J369v09n02_01

A Theoretical Framework for Multi-Unit Management Development in the 21st Century

Christopher C. Muller
Robin B. DiPietro

SUMMARY. The current study proposes a theoretical framework for the development of multi-unit managers. Chain restaurants that use multi-unit managers as a part of the infrastructure of their operations have been a growing segment of the foodservice industry for the past half century (Muller & Woods, 1994). This segment of managers is growing as the 21st century heralds a continued increase in the number of multi-unit restaurant organizations and thus an increase in the need for numbers of skilled multi-unit operators. The study proposes a model for the development of multi-unit managers through a series of phases that will help develop characteristics for success. It then further proposes a framework for training and development of managers in order to fill the increasing need for multi-unit managers. Implications for practitioners

---

Christopher C. Muller, PhD, is Professor, Rosen College of Hospitality Management, University of Central Florida, 9907 Universal Blvd., Orlando, FL 32819 (E-mail: cmuller@mail.ucf.edu).

Robin B. DiPietro, PhD, is Assistant Professor, Rosen College of Hospitality Management, University of Central Florida, 9907 Universal Blvd., Orlando, FL 32819 (E-mail: dipietro@mail.ucf.edu).

[Haworth co-indexing entry note]: "A Theoretical Framework for Multi-Unit Management Development in the 21st Century." Muller, Christopher C. and Robin B. DiPietro. Co-published simultaneously in *Journal of Foodservice Business Research* (The Haworth Hospitality & Tourism Press, an imprint of The Haworth Press, Inc.) Vol. 9, No. 2/3, 2006, pp. 7-25; and: *Human Resources in the Foodservice Industry: Organizational Behavior Management Approaches* (ed: Dennis Reynolds, and Karthik Namasivayam) The Haworth Hospitality & Tourism Press, an imprint of The Haworth Press, 2006, pp. 7-25. Single or multiple copies of this article are available for a fee from The Haworth Document Delivery Service [1-800-HAWORTH, 9:00 a.m. - 5:00 p.m. (EST). E-mail address: docdelivery@haworthpress.com].

Available online at http://jfbr.haworthpress.com

doi:10.1300/J369v09n02_02

are presented along with a call for further research in this understudied area.
doi:10.1300/J369v09n02_02

KEYWORDS. Multi-unit restaurant management, management training and development, key success characteristics

INTRODUCTION

Multi-unit restaurants or "chain" restaurants represent a large portion of the U.S. foodservice industry. Over the past 50 years, the restaurant industry has grown to be a staple of the American diet, and large national chain restaurants lead the pack when it comes to growth and expansion. Overall, the foodservice industry represents $476 billion in revenue and over 4% of the gross domestic product of the U.S. Over 50% of the total annual domestic restaurant sales are accounted for by the leading chain restaurant organizations (Muller & Woods, 1994; National Restaurant Association, 2005). There are over 900,000 restaurant units, with approximately 270,000 of them representing chain restaurants (National Restaurant Association, 2005). With an estimated span of control of 7 restaurants per multi-unit operator, this means that there are approximately 38,571 multi-unit managers working in chain restaurants today. While the U.S. foodservice industry is growing in revenues at a rate of approximately 4% annually, the top 25 chain restaurants grew approximately 5.1% in total revenues during the same time period (National Restaurant Association, 2005; Technomics, Inc., 2004).

This continued growth and the corresponding organizational complexity it brings means there is a distinct opening in the market for well-trained and qualified multi-unit managers. These are the professional knowledge workers who Peter Drucker called the "manager of managers" (Drucker, 1955, p. 24). There is a need for a set of organized and empirically tested key success factors that multi-unit managers can apply in order to be successful in their role. This brings an additional requirement for the implementation of a training and development agenda for chain restaurant organizations in general and multi-unit managers specifically. The theoretical framework we propose creates an agenda and a training rubric for multi-unit management development in the restaurant industry.

BACKGROUND LITERATURE

Single-unit managers require a very different skill set (including the success factors that are part of their job set) than do multi-unit managers. Reynolds (2000) found that the top five characteristics required of a single unit manager were: organizational skills; interpersonal skills; restaurant experience, knowledge and skills; honesty, integrity and strong ethics; and leadership skills. This research was done through interviews with industry leaders in top chain restaurants and the skills were found to be similar to key success factors required of managers in a variety of other business organizations. Despite the small sample size of the Reynolds (2000) study, these key characteristics have been found in other studies as well (Kakabadse & Margerison, 1988; Boulgarides & Rowe, 1983; Van der Merwe, 1978).

Research regarding the attributes and activities that multi-unit managers perform has been published over the past twenty years by Umbreit and Tomlin (1986), Umbreit and Smith (1991), Muller and Campbell (1995), and Umbreit (1989, 2001). These studies suggested that the knowledge, skills and behaviors of the multi-unit restaurant manager are measurably different from single unit restaurant managers (Umbreit & Tomlin, 1986; Umbreit, 1989).

In the initial research into the role and responsibilities of the multi-unit manager, Umbreit (1989) interviewed executives and practicing multi-unit managers to determine what the key job aspects were for a manager of two or more restaurants. The results suggested that restaurant operations were the most important job aspect of the multi-unit manager, followed closely by human resources management. The respondents in this study indicated that they spent over half of their time on the job performing tasks related to problems in restaurant operations and resolving issues in human resources.

Umbreit found that some of the self-determined key success factors that multi-unit managers said they needed prior to being promoted to their new positions included: how to manage managers, how to motivate people, how to work with a diverse group of people, how to get things done and solve problems through other people, how to build teams, how to manage time and set priorities, how to delegate, how to deal with unstructured time, how to enforce standards in multiple restaurant units, and how to recognize differences in each unit's operational situation (Umbreit, 1989; Umbreit & Smith, 1991).

A follow-up study done by Muller & Campbell (1995) intended to further the Umbreit findings by expanding the research to a large empirical study with a single quick service restaurant chain. This study built on the results that Umbreit (1989) drew on from the smaller sample sizes in previous studies. The research paper analyzed the differences in perceptions of the single unit (store or restaurant) managers, district managers, and headquarters or corporate staff personnel regarding the tasks or skills of the multi-unit manager. The results determined that all of the respondents believed that they were competent at their own positions, but did not feel as competent at the thought of being promoted to the next level of supervision. Store managers especially felt that they needed more training in marketing and in human resources to move up to the district level. The authors suggested this emphasized the store manager perception that the mastery of human resources management was a crucial component of the job of a multi-unit manager.

In their research Umbreit and Tomlin (1986) proposed that there are five overarching key dimensions of a multi-unit manager's job–restaurant operations, marketing, finance, facilities management and human resource management. Muller and Campbell (1995) replicated these dimensions in their study and had participants rank the importance of each of these dimensions to job success. A key finding in this article was that all levels of management were in consensus regarding the rank order of the three most important job dimensions of a multi-unit manager. Of the five dimensions, the most important was human resources, followed by restaurant operations and then finance (Campbell, 1994; Muller & Campbell, 1995). These findings are used as the basis for the creation of an agenda for the development of promotable unit managers into successful multi-unit managers.

Unfortunately, there have been few studies since then that have further investigated the role of the multi-unit manager in restaurant operations. Umbreit (2001) published a qualitative Delphi-style study of ten chain restaurant executives who were asked to review and comment on the changing role of the multi-unit manager. The findings of this article suggested that the expanding span of control for multi-unit managers seen in the late 1980's and early 1990's had reverted back to the tighter spans of control of previous decades. One additional outcome of this study was the suggestion from the executives that titles and specific responsibilities had changed for multi-unit managers. In their view the district management role in 2001 required more "soft skills" as compared to the more traditional "hard" or technical skills. This finding re-emphasizes the importance of the human resource skills that previous studies had found to be a critical component for the success of the multi-unit manager. This perspective concluded that to be successful in a time of labor and skill shortages there would be an increasing importance placed on "new" multi-unit managers to possess excellent communication, team building and motivating skills. This research emphasized the need for continued research into the changing roles and development needs of multi-unit managers.

The existing literature shows a gap in the application of the research regarding a formalized development program for multi-unit restaurant managers and the need for targeting the key success factors found in the research. This proposed theoretical framework will develop an outcome based multi-unit manager development program to help close this gap.

A Model of the "Phases" of Multi-Unit (District) Manager Development

Taking advantage of the seminal work done in the identification of the attributes and role dimensions of the multi-unit restaurant manager cited above, a model (see Figure 1) was developed to show the hierarchy of manager development. The highlights of this model include the path from being a "Super Operator" with a heavy reliance on the technical skills of the organization to the mastery of the more complex management issues of human resource development.

FIGURE 1. "Phases" in Multi-Unit (District) Manager Development

As shown in the model, a developmental training program for the newly promoted district manager would be constructed on an assumption of personal growth over a period of time. Upon promotion from a successful single unit General Manager position, the new district manager would most likely exhibit the attributes and behaviors associated with a "comfort zone" of existing skills. These skills would include the dimensions of Restaurant Operations and Facilities Management and would by necessity have a tactical short-term decision window. Basically a general manager's measurement of success might include (but not be limited to) unit-level tasks such as scheduling, cost control, building maintenance and customer satisfaction. Left on their own, the best candidates will discover the need for new skills and practices. Alternatively, a personal development program could be used to advance this process, to the benefit of both the district manager and the company which employs him/her.

Using the model, as the new district manager gains experience and perspective, a realization should set in that the Finance dimension is more complex and mastery requires new skills and concepts. To drive district level profits a better understanding of forecasting, budgeting and cash flows is a prerequisite. It also becomes apparent that to budget for preventive maintenance or capital investment in physical plant equipment and repairs requires a longer decision making time frame that needs to be developed in the multi-unit manager. Again, a theoretical personal development plan could expedite this realization and advance the time spent in on-the-job training.

To facilitate district growth, a new appreciation for the role of the Marketing dimension should appear in the multi-unit manager. Marketing is less skill based and more conceptual, focusing on brand management, the role of competition in the marketplace, and such issues as pricing and positioning of the district's restaurants. It is reasonable to expect that the experienced multi-unit manager's decision time frame is extended from short-term to mid-term. Implementation of successful district marketing campaigns may require 6 to 9 months of planning and development. As in the other three dimensions, a fully defined training program could address these issues.

The final dimension is the one which deals with the very "soft" issues of Human Resources. No district manager can be fully empowered to control their complement of restaurants until they begin to see that their role has evolved from "task master" to "people developer." Restaurant companies with district managers who understand the Human Resource function should expect to see lower staff turnover rates–better yet they should see higher levels of retention of key employees as commitment grows.

The focus for such individuals is to build leadership, high performance teams, and collaboration amongst staff at the restaurant unit level. A good development program would include all the things mentioned above, plus an appreciation of how different people learn, are motivated to succeed, and at what pace they develop. While the model suggests the path for development could take up to two calendar years, each newly promoted or newly hired multi-unit manager would proceed at a self-directed pace.

THEORETICAL FRAMEWORK FOR DEVELOPMENT AND TRAINING

The proposed framework is to create an outcome based multi-unit manager development program that will highlight the key underpinnings of multi-unit management development and the key success factors that make multi-unit managers different from their single unit management counterparts. Through application of empirical studies and practitioner based research that has been completed on multi-unit management in the restaurant industry, the "phases" model (Figure 1) will be used to identify the needs of the multi-unit restaurant industry and develop a theoretical framework for a development program for multi-unit restaurant managers. This framework will allow the researchers to develop, by using the key success factors determined through past research, a training program for multi-unit managers in order to help the industry to develop successful skills in the large number of multi-unit managers in the restaurant industry.

The multi-unit management development program should be comprised of a number of activities that will help the new multi-unit manager learn and develop new skills and understand how their new job is comprised of different key success characteristics and is more encompassing than their previous job as a single unit manager. The proposed outline for an outcomes based training program for new multi-unit managers follows.

Phase I–On-The-Job Training (Purpose is to develop multi-unit manager transition skills–moving from single unit manager to multi-unit manager).

A. Position Orientation

1. Position Orientation/Description (Multi-unit management job description, accountabilities and tasks involved in successful performance of the job)

2. Office/Administration Orientation (Orientation with each of the areas in the administrative office)
3. District Restaurant Orientation

B. District Restaurants Initial Assessment

1. Operations Performance and Assessment
2. Management Skills Assessment
3. Crew Skills and Assessment
4. Profitability–Cost Controls and Sales Building Assessment

C. Identify, Develop and Implement Goals and Supporting Programs for each Restaurant
D. Develop MUM Monthly Work Schedule
E. Complete Program Reading Assignments

Phase I–Outcomes: MUM will develop skills in: (1) Assessing restaurant operations (2) Accurately identifying restaurant goals and programs to support assessments (3) Identifying high potential individuals at the unit level (4) Scheduling and time management.

When a Multi-Unit manager completes Phase I of training they will have:

1. Completed the administrative "set-up" for the district, which will include:

   A computer and paper filing system
   A system for analyzing restaurant P&Ls and reporting on performance and
      Explaining required action steps
   Setup a district communications system
      Emails
      Voicemails
      Fax
   Setup and/or review with the Restaurant Managers:
      Emergency actions procedures
      Reporting operating statistics

2. Completed (or coached RMs to complete) Personal Development plans for each manager in their district. (Within three weeks

of completion, each manager will have had an appraisal discussion and written development goals for the coming 6 months.)

RMs discuss development plans with their AMs
MUM discuss development plans with each RM in the district.

3. MUM will edit (or write) their Job Description and schedule a time to review it with their Supervisor.
4. If the new MUM started with 3 or 4 restaurants; their district will be increased to 6 Restaurants.
5. Develop a 6 month district development plan for their restaurants. (This should be done about 6 months after the MUM is in their job.) This plan should include activities that:

   Improve their districts' operating performance
   Improve their districts' financial performance
   Include marketing and sales promotions activities
   Successfully develop restaurant management teams
   Increase retention rates across all levels.

6. Draft a Personal Development plan for themselves and review it with their supervisor
7. Determine development levels for each manager in their district (based on development levels for situational leadership)
8. Work an open and close with each restaurant crew at least once each quarter.
9. Identify, develop and prioritize goals for each restaurant in the district. Include the restaurant management team in this process–the teams' development level would play a part in defining the process.
10. Develop a district operations, marketing and financial plan.
11. Write a monthly MUM work schedule that will allow the MUM to meet district monthly goals.
12. Recognize and talk with frequent customers in each of the district restaurants
13. Learn sales patterns and customer flow in each district restaurant.
14. Develop a self understanding of their stress levels and learn various body signals that warn them they need to break because their decisions may not appropriate for the situation.
15. Teach their managers how to identify and prevent approaching operating problems. Based on cause and effect reasoning managers

should be able to see behaviors that if continued would cause particular problems (e.g., manager working D/T during meal periods lead to poor service, messy dining room, loss of customers, reduced sale, lower profits.)

Once a multi-unit manager has completed Phase I, he/she will have the ability to accomplish results through their team of restaurant managers. There will have been measurable changes in the multi-unit management skills for the participant in the areas of:

- Delegation
- Planning and monitoring
- Time management
- Unit manager development
- Communication
- Persuasion and motivation

Phase II–Classroom Training (Purpose: Develop MUM district management performance and improve district restaurants operating and financial results.) Phase 2 primarily takes place out of the restaurant environment and continues to build on the multi-unit manager's skill set. By giving the multi-unit manager additional information that will help them grow in their knowledge, it will help the application of the knowledge in the restaurants and on the job. The goal of Phase 2 will be to help the personal development of the multi-unit manager and increase the link between book knowledge and actual application of that on the job to help the district improve operational and financial performance.

The Phase 2 training will involve developing teamwork and team based training. The teams of multi-unit managers will be able to learn a competency, discuss through the role play of various situations and discuss the application of the competency through discussions of actual situations that have occurred in the various restaurants. This phase of training will be a combination of applied theory and training modules.

Some of the theoretical topics covered will be: strategic issues for multi-unit managers; multi-unit restaurant life cycles; development of a foodservice manager; key success factors for a multi-unit manager. The training topics will include: "The 60 Day Plan: Achieving Maximum Results"; Building Your Own Position Descriptions; Situational Leadership; Giving and Receiving Feedback; Managing Your District's Financial Performance; Conducting Effective District Meetings.

Throughout Phase 2, the multi-unit manager will be working on his/her Personal Development Workbook that will be used throughout the multi-unit manager career. This book will give structure to the job of a multi-unit manager.

A. Human Resources and Operations

1. Complete a district tactical plan for the next 6-9 months and brief Supervisor and/or owner.
2. Lead District in implementing a new project (e.g., product introduction, restaurant opening)
3. Manager Development (work with the various managers in their district to help develop them. It is important for the MUM to get exposure to all of the facets of management development in order to assist the restaurant manager with this area of operations)

   3.1 Manager Trainee
   3.2 Night Shift Managers
   3.3 Restaurant Managers Development

4. District Communications

   4.1 District Meetings–Evolve Format
   4.2 Restaurant Crew "Round Table" meeting (with RM)
   4.3 District "Tours" with RMs

B. District Financial Results

1. Monthly Budget Presentation
2. Annual Budget Preparation

   2.1 Sales Goals
   2.2 Expenses Goals
   2.3 Operations and Marketing Programs to Support Budget Results

C. Marketing, Promotions and Public Relations

1. Develop restaurant management's awareness of customers preferences
2. Train Restaurant Managers to assess competitors operations, advertising & promotions to determine how they can successfully compete

3. Develop local store marketing plans for each restaurant
4. Attend regional/ADI marketing meetings.

D. Facilities and Risk Management

1. Develop and/or direct a district Safety & Risk Management Committee
2. Ensure district restaurants comply with legal requirement and governmental regulations.

Phase II–Outcomes: MUM's district results (after completing Phase II Classroom Training) will show:

1. Improved district operating performance
2. Improved district financial performance
3. Successfully developed unit management teams
4. Increased retention rates across all labor levels
5. Improved understanding of all of the "phases" of the multi-unit manager development model

Through Phase 2, the new multi-unit manager will learn the capabilities of the restaurant management teams in their district.

1. Completed plans for improving each district restaurant's operating and financial performance.
2. Conducted Restaurant Manager meeting for his district using the following outline.
3. Start Restaurant Manager District tours. MUM and all RMs ride together to visit and evaluate each restaurant.
4. Change RMs district meeting format and leadership–RMs plan and lead district meetings. The MUM helps with preparation and attends the meeting, but the RM conducts the meeting.
5. The MUM starts conducting regular AM meetings for the district.
6. The MUM helps each RM plan to conduct a crewmember "round table" where crew members from each restaurant in the district meet to discuss Customer Service and how to improve it.
7. The MUM spends each new manager Trainees first training day working stations with the Trainee.
8. The MUM spends 50% of restaurant workdays on the night shift training the Night Manager and night crew (including proper closing procedures and inventory).

Phase III–Linking On-The-Job With Classroom Training. The purpose of the last component of the theoretical model for training multi-unit managers is to help multi-unit managers with the link between the training and development of their skills and the actual performance on the job. In order to ensure that the MUM is developing and assessing all of the components of the restaurants in the district, a formal review of each restaurant should be undertaken at least once a year. Here is a sample outline of a District Assessment Form that should be used to evaluate the individual units.

District Assessment Form

Complete an Annual Assessment and Improvement Plan for their district. This Plan would include detailed information reflecting the status of:

I. Company Management
   - Vision Statement
   - Mission Statement
   - Core Company Values

II. District Management
   - District Communications Systems
   - District Control Systems
     - Computer System
     - P&L Reports
     - Labor Formula & Staffing Guides
     - Production Level Systems
     - Customer Service Systems

III. Risk Management
   - Safety Audits
   - Emergency Procedures
   - Legal Compliance
   - Fire Emergency Plan
   - Hot Water Heater
   - Restaurant Security

IV. Operations
   - Operations Manual
   - Sales Volumes
   - Average Check
   - Key menu Prices

- Customer Count
- D-T Car Count
- Sales Percentages
  - Customer Contact Areas
  - Day Parts
- Speed of Service
  - Counter
  - D/T
- Labor Statistics
- Controllable Expenses
- Non-controllable Expenses
- Inventory Percentages

V. Marketing Program
- Local Store Marketing
- Premium (Gift) Promotions
- Sales Promotions
- Community Relations
- Promotion Payback Measurements

VI. People
- Restaurant Supervision
- Employee Retention
- Training Systems
  - Orientation Program
  - Crew Training System
- Performance Appraisal Programs
- Hiring Systems
  - Crew
  - Manager
- Company Training Programs
- Time & Attendance Records
- Pay Roll Policies and Procedures
- Promotion Systems
  - Crew
  - Management

VII. Human Resources
- Time & Attendance Reports
- Employee Handbook
- Policy Manual
- Required Posters
- Internal Communications System
- Employee Personnel Files

Legal Compliance Records
Job Descriptions
Pay Practices
Employee Benefits
Performance Management Systems
Workers' Comp Experiences
Key Company Policies

VIII. Site & Restaurant Facility Parking Lot
Curbing
Walkways
Patio
Lighting
Trash Enclosures
Landscaping
Playground
Pole and Building Signage
Drive-Thru Area

Building Exterior
Roof Condition
Exterior Finish
Doors
Roof
HVAC

Building Interior
Entrance and Dining Room Walls
Front Counter
Drink Service Area
Dining Room
Menu Board
Floors
Ceilings
Restrooms
Kitchen & Office Area
Crew Room
Storage Areas
Utility Area
Roof Access
Plumbing

Equipment
Fryers
Fry Computers

Ice Machine
Drink System
Broilers/Grills/Ovens
Food Holding Cabinets
Circuit Breaker Box
Other Equipment
  Fire Extinguishing System & Extinguishers
  Heated Equipment
  Refrigerated & Frozen Equipment
  Hot Water Heater

IX. Restaurant Operations
Food Safety
Operating Systems
Product Quality
Guest Service

Throughout the training program, in order to help the managers assimilate the skills that they are being taught, there will be hands on activities and the use of case studies in order to ensure that they are transferring skills to real world situations. This proposed training framework lays the foundation for how to develop single unit managers into multi-unit managers using the basic theoretical framework reviewed in the "phases" model for multi-unit management development and puts it in a proposed framework that will help to ensure that all of the skills needed are developed in new and even in existing multi-unit managers. It is not enough to just hope that good single unit managers develop into multi-unit managers. Organizations need to focus on actually developing the strengths needed of multi-unit managers and working with these managers on a consistent basis to improve the skill set of people.

IMPLICATIONS FOR PRACTITIONERS

The current study proposes a model of multi-unit management development that is based on a theoretical research foundation. It takes the research that has previously been completed regarding key success factors of multi-unit managers and puts it in a model that can help practitioners visualize the development of people in the positions of multi-unit managers. In a continued time of labor and skill shortages in the restaurant industry today, there will continue to be an increasing importance placed on "new" multi-unit managers possessing excellent communication, team building and motivating skills.

The development and training framework that is presented is an outline for a development program for new or existing multi-unit managers that can be used by industry practitioners looking to develop certain skills in multi-unit managers. It takes the conceptual model for multi-unit management development and puts it into a format that can be delivered and used by multi-unit managers to grow their skill set and help with improved performance of chain restaurant organizations. Practitioners need to constantly look for ways to develop "soft" skills in multi-unit managers and this program works with a combination of on-the-job learning and development and with theory and classroom delivery of information to try to get the best out of people.

CONCLUSION

Single unit managers have very different areas that require their attention in managing their one restaurant compared to multi-unit managers who are managing many. District managers have to focus on a larger span of supervisory control and ensure that things in a district run smoothly versus just concentrating on a single unit operation (Reynolds, 2000; Muller & Campbell, 1995). There has been research done regarding the characteristics of multi-unit managers, but there has been a gap in presenting a pragmatic development strategy to develop those multi-unit characteristics in people. Training for single-unit managers is done on a regular basis in organizations, but the skill set that they need to have is not as complex as that of a multi-unit operator. In an attempt to fill the gap in the research and development programming for multi-unit managers, the current study takes the findings of researchers regarding these skills and puts them into a framework that can be applied to any restaurant organization with multiple units.

Despite the large number of multi-unit restaurant managers in the U.S., the authors have not found a proposed training and development program based on past research in the area. The current framework for training has been proposed using the key success factors and characteristics from previous research and developed into the "Phases" in Multi-Unit (District) Manager Development (Umbreit, 1989; Umbreit & Smith, 1991; Muller & Campbell, 1995).

This proposed theory-based framework creates an outcome focused development program. When this framework is applied it will assist complex restaurant organizations in their efforts to design effective training and development planning. In turn, this will enhance the work of multi-unit operators by focusing on key characteristics and components for success.

REFERENCES

Boulgarides, J. D., & Rowe, A. J. (1983). Success patterns for women managers. *Business Forum, 8*(2), 22-24.

Campbell, D. F. (1994). *Critical Skills for Multi-Unit Restaurant Management*. Unpublished Monograph, Master's Thesis.

Drucker, P. (1955). *The Practice of Management*. London: Pan Books.

Kakabadse, A., & Margerison, C. (1988). Top executives: Addressing their management development needs. *Leadership & Organization Development Journal, 9*(4), 17-21.

Lefever, M. M. (1989). Multi-unit management: Working your way up the corporate ladder. *The Cornell Hotel and Restaurant Administration Quarterly, 30*(1), 61-67.

Muller, C. C. (1999). The business of restaurants: 2001 and beyond. *International Journal of Hospitality Management, 18*(4), 401-413.

Muller, C. C., & Campbell, D. F. (1995). The attributes and attitudes of multiunit managers in a national quick-service restaurant firm. *Hospitality Research Journal, 19*(2), 3-19.

Muller, C. C., & Woods, R. H. (1994). An expanded restaurant typology. *The Cornell Hotel and Restaurant Administration Quarterly, 35*(3), 27-37.

National Restaurant Association (2005). Restaurant Industry 2005 Fact Sheet. Retrieved on May 5, 2005 from http://www.restaurant.or/pdfs/research/2005factsheet.pdf.

Paul, R. N. (1994). Status and outlook of the chain-restaurant industry. *The Cornell Hotel and Restaurant Administration Quarterly, 35*(3), 23-25.

Reynolds, D. (2000). An exploratory investigation into behaviorally based success characteristics of foodservice managers. *Journal of Hospitality & Tourism Research, 24*(1).

Ritchie, B., & Riley, M. (2004). The role of the multi-unit manager within the strategy and structure relationship; evidence from the unexpected. *International Journal of Hospitality Management, 23*(2), 145-161.

Technomics, Inc. (2004). 2004 Technomic Top 100 Report. Retrieved on May 5, 2005 from http://www.technomic.com/facts/top_100.html.

Umbreit, W. T. (1989). Multiunit management: Managing at a distance. *The Cornell Hotel and Restaurant Administration Quarterly, 30*, 53-59.

Umbreit, W. T. (2001). Study of the changing role of multi-unit managers in quick service restaurant segment. In H. G. Parsa & F. A. Kwansa (Eds.), *Quick Service Restaurants, Franchising and Multi-Unit Chain Management* (pp. 225-238). New York: The Haworth Hospitality Press.

Umbreit, W. T. & Smith, D. I. (1991). A study of the opinions and practices of successful multiunit fast service restaurant managers. *The Hospitality Research Journal, 14*, 451-458.

Umbreit, W. T., & Tomlin, J. W. (1986). Identifying and validating the job dimensions and task activities of multi-unit foodservice managers. *Proceedings of the 40th Annual Conference on Hotel, Restaurant, and Institutional Education*, August, 1986, pp. 66-72.

Van der Merwe, S. (1978). What personal attributes it takes to make it in management. *Ivey Business Quarterly, 43*(4), 28-32.

doi:10.1300/J369v09n02_02

Managing Language Policies in the Foodservice Workplace: A Review of Law and EEOC Guidelines

Bonnie Farber Canziani

SUMMARY. The principal objective of this paper is to review the business and law literatures to determine the legal, business and ethical issues at stake in setting language policies for personnel in foodservice businesses within the United States. Furthermore, we review the types of language policies and practices that have resulted in claims of employee discrimination under Title VII of the U.S. Civil Rights Code, and where applicable, we note specific legal rulings that have relevance for future managerial decision-making and future research in the foodservice industry. We additionally identify practices for managing a linguistically diverse workforce that are seen to be both beneficial and compliant with existing civil rights law. doi:10.1300/J369v09n02_03 *[Article copies available for a fee from The Haworth Document Delivery Service: 1-800-HAWORTH. E-mail address: <docdelivery@haworthpress.com> Website: <http://www.HaworthPress.com> © 2006 by The Haworth Press, Inc. All rights reserved.]*

Bonnie Farber Canziani is Director, Hospitality and Tourism Management, Department of Recreation, Tourism and Hospitality Management, University of North Carolina at Greensboro, P.O. 26170, Greensboro, NC 27402-6170 (E-mail: bonnie_canziani@uncg.edu).

[Haworth co-indexing entry note]: "Managing Language Policies in the Foodservice Workplace: A Review of Law and EEOC Guidelines." Canziani, Bonnie Farber. Co-published simultaneously in *Journal of Foodservice Business Research* (The Haworth Hospitality & Tourism Press, an imprint of The Haworth Press, Inc.) Vol. 9, No. 2/3, 2006, pp. 27-47; and: *Human Resources in the Foodservice Industry: Organizational Behavior Management Approaches* (ed: Dennis Reynolds, and Karthik Namasivayam) The Haworth Hospitality & Tourism Press, an imprint of The Haworth Press, 2006, pp. 27-47. Single or multiple copies of this article are available for a fee from The Haworth Document Delivery Service [1-800-HAWORTH, 9:00 a.m. - 5:00 p.m. (EST). E-mail address: docdelivery@haworthpress.com].

Available online at http://jfbr.haworthpress.com

doi:10.1300/J369v09n02_03

KEYWORDS. Civil rights, employment law, bilingualism, language policy, English-only, EEOC

OVERVIEW OF THE SITUATION

Immigration is a continuing reality for the United States, impacting the profile of the nation's workforce and increasing the number of candidates for employment in the foodservice sector for whom English is a second language. The U.S. Census Bureau reports that approximately 47 million people over the age of five (18 percent of all U.S. residents) speak a language other than English at home. These data represent an increase from 14 percent (31.8 million) in 1990 and 11 percent (23.1 million) in 1980. A total of 21,320,407 residents indicate that they speak English less than "Very well" (Shin & Bruno, 2003). The Census Bureau estimated the undocumented population at 8 million in 2000. Because it also estimated a net annual increase of 450,000, an updated Census estimate would be closer to 11 million in 2005 (Stern, 2005).

Of particular note is the fact that the restaurant industry has a disproportionate number of workers who are hired with limited English-speaking skills (Jackson, 2002). According to the U.S. Department of Labor's Bureau of Labor Statistics, 12 percent of foodservice employees are foreign-born compared to 8 percent for all other occupations. The restaurant industry is the single largest employer of immigrants in the nation. More than 1.4 million of the restaurant industry's 8 million employees are legal immigrants. Presumably more are working in restaurants unlawfully (Lydecker, 1996). This creates a work environment where many non-English languages are spoken on the job, including Spanish, Chinese and Vietnamese (Jackson, 2002). The language problem is rife with frustrations for employers, employees, and others impacted by the communication gaps that arise when people do not share a common language. The impacts are multiple, affecting individual workers, social classes, organizations, and the country as a whole. From a monetary perspective, this plurilingualism can have a negative economic impact on the foodservice industry. According to economists, limited English skills of foreign-born U.S. workers cost "$65 billion annually in lost productivity" (Bahls & Bahls, 1998).

There are several concerns when studying language issues in relation to U.S. business sectors like the foodservice industry. The first concern is to what kinds of language policies tend to be implemented in foodservice operations. Another concern is to examine the perceived divergence between (a) the employer's right to set language policies or requirements in the workplace (including English-only policies) to satisfy business necessities and (b) the preservation of the civil rights of employees and affected others, e.g., non-English speaking customers. In 2002, the Equal Employment Opportunity Commission received 228 charges challenging English-only policies in the workplace, which is a cautionary note to any manager in the foodservice sector (U.S. Equal Employment Opportunity Commission, 2002).

LANGUAGE POLICIES IN THE U.S. WORKPLACE

Language policies in the workplace can range from informal practices to formal written policies and consequences. Some of these policies refer to setting narrow restrictions on what language(s) can be used in the workplace; other policies refer to the requirement of certain languages for critical employee personnel decisions such as hiring, performance appraisal and promotion. Two primary governing bodies have offered interpretations of what business necessity actually means in the context of language policies and civil rights: (1) the Equal Employment Opportunity Commission (EEOC) and (2) various levels of the judicial system. The former agency monitors the application of Title VII of the Civil Rights code in business decisions related to recruitment, hiring, promotion, transfer, wages and benefits, work assignments, leave, training and apprenticeship programs, discipline, and layoff and termination in companies of 15 or more employees. We note that the Department of Justice's Office of Special Counsel for Immigration Related Unfair Employment Practices (OSC) is responsible for investigating charges of job discrimination related to an individual's language or national origin in workplaces with 4 to 14 employees. The courts come into play primarily when claims of discrimination due to national origin have made their way up the judicial ladder; they have up to now rendered mixed verdicts on the language policies employers are implementing in the U.S. workplace.

The EEOC has issued specific guidelines under its documentation on national origin discrimination, which are publicly available to employers contemplating language policies: "Employment discrimination against a national origin group includes discrimination based on physical, linguistic, or cultural traits closely associated with a national origin group (U.S. Equal Employment Opportunity Commission, 2002)." The

Commission further marks language-related policies as potentially suspect in the following warning:

> Employers sometimes have legitimate business reasons for basing employment decisions on linguistic characteristics. However, linguistic characteristics are closely associated with national origin. Therefore, employers should ensure that the business reason for reliance on a linguistic characteristic justifies any burdens placed on individuals because of their national origin. The subsections below provide guidance on employment decisions that are based on foreign accent or fluency, and guidance on policies requiring employees to speak only English while in the workplace (U.S. Equal Employment Opportunity Commission).

Language as a Bona Fide Occupational Qualification

Among business employers in the United States, language ability, particularly in spoken and written English, has become more and more a preferred occupational qualification and assigned value in the way industry generally assigns worth to technological or other industrial skills. Estimating a value for language skill is not an entirely new concept, since language teachers, interpreters, and other candidates for language-centered jobs have traditionally been asked for some credential proving linguistic expertise. Today, however, in more and more business sectors faced with the realities of globalization in the labor force, language skills are increasingly linked to an individual's opportunity to advance in the U.S. workplace. The issue of language can impact all phases of an employee's interaction with a potential or existing employer. Most problems occur when businesses whose core service is not identified as a language product, i.e., language teaching or translating, try to justify language requirements as business necessities for employment decisions.

When discrimination claims arise, the courts' overriding concern is whether a plaintiff claiming discrimination against an employer can prove a prima facie case of (a) disparate treatment and/or (b) disparate impact as a member of a national origin class protected under Title VII. In disparate treatment cases, employees claim that an employer deliberately discriminated against a protected class on the basis of national origin by setting a specific language policy or requirement. An allegation of disparate impact is one where employment practices that appear to be neutral are said to yield disproportionately greater negative consequences for a protected minority group; resulting impacts may be considered without regard to employer motivation or intention to discriminate. At this point, we turn to a review of civil rights cases claiming discrimination in various personnel practices (for the most part based on national origin) to evaluate the current stance on language policies relevant to foodservice managers in the U.S.

Staffing Decisions

When foodservice managers seek to staff job positions, it may be the case that they consider advertising for or selecting candidates who speak specific languages, for example, posting an ad in a classified section that states 'Must be bilingual.' As indicated previously, the EEOC will examine carefully any employer's denial of a job opportunity to an applicant because of the applicant's foreign tongue or accent. On the other hand, the ability to communicate is clearly linked to many jobs, and the courts have been lenient in perceiving the employer to be entitled to require a reasonable level of English proficiency where the job demands it.

The courts have ascertained that business necessity can be established by showing that assigned language-related duties hold a reasonable relationship to the specific occupational job and are essential to perform the job in a reasonable manner (*Garcia vs. Rush-Presbyterian St. Luke's Medical Center*, 1981). A test of business necessity would require a foodservice employer to ask if a language requirement is legitimately needed based on the nature of the occupation or its primary tasks, e.g., requiring reasonable English fluency for a phone clerk handling pizza take-out orders, when the take-out clientele is statistically proven to be predominantly English-speaking. On a related point, *Vasquez vs. McAllen Bag & Supply Co.* (1981) upheld English as a hiring requirement, citing in part the business advantage and convenience of having truck drivers able to understand instructions from the owners, who spoke only English.

However, the need for English fluency is highly dependent on demonstrable frequency of public contact in the job or of the need to give or receive instructions in English or to perform other oral and written tasks that are expected by a foodservice operation to be conducted in English, e.g., preparing nightly menu boards or inserts. Above all, employers should avoid having a blanket language fluency requirement. Each job requiring language fluency must be scrutinized on a case by case basis for business necessity, and clear task-to-fluency justifications must be documented. It has been noted that:

> The EEOC has taken the stand that some positions, including those typical of manual labor and production jobs, can be performed satisfactorily with little or no knowledge of English. A suit filed by the EEOC against the Sheraton National Hotel in Virginia involved a dishwasher who was temporarily laid off in September 2001 while the hotel remodeled its restaurant. Employees were advised they would be rehired less than a year later, the suit said. However, Romero, who worked for the hotel for 16 years, was denied the job because of a newly implemented English fluency requirement, the commission said. (Joyce, 2004)

EEOC acknowledges that an individual may have sufficient fluency in English to function well as a cook in a fast food restaurant, but not have sufficient written English skills to successfully work as the manager of that operation, performing periodic operational analyses and reports.

With respect to requiring job candidates to take candidacy or certification tests in English, as long as the employer applies this selection device to all candidates and the test demonstrates content validity (Burns, 1995, 1996) in the context of the occupational position or classification, the employer is within his/her rights. The court in *Frontera vs. Sindell* (1975) permitted an English civil service exam for carpenters, stating that the technical terms in the test, while indisputably in English, were such that any trained carpenter wishing to work within the United States would be expected to recognize and utilize them. It is understood, however, that tests can not be employed exclusively to eliminate applicants of a particular national origin.

One legal gray area in the job placement of a candidate or current employee is how the courts view a business rationale based on supposed customer preference for English-speaking or non-accented individuals. Case decisions relying on a defense of customer preference have for the most part occurred outside the realm of national origin discrimination claims, e.g., Hooters of America, Inc., which has been permitted to make gender-based hiring decisions to preserve their product concept. To copy their argument, foodservice operators would have provide concrete evidence that using employees of the non-preferred language will cause the essential nature of the product to be significantly altered from what the preferred-language employee delivers.

Smith sides with the non-English speaker in this debate:

> issues of fundamental fairness come into play when people are virtually forced to give up their mother-tongue to conform to employer and customer expectations of what an individual's English should sound like.... Instead of lay persons' preferences, the focus should be on the evaluations of linguistic experts in regards to the plaintiff's speech. In addition, I believe this assertion is supported by the findings of linguistic research in the area of listener prejudice. (2005)

Employee Advancement and Compensation Decisions

In general, employers are allowed to make initial and future job assignments based on language ability as long as there is a verifiable and legitimate business necessity. There is, however, divided opinion in the case law where a business firm knowingly hires a worker that has limited or no proficiency in a required language, and then subsequently considers that linguistic deficiency as a factor in critical personnel decisions related to that job. A number of Title VII cases have gone to court contesting employers' promotion and compensation decisions. Cases where the verdict was in favor of linguistic minority plaintiffs include an employee who was well-qualified but not promoted to higher-level jobs because of her Polish accent (*Berke vs. Ohio Dept. of Public Welfare*, 1996) and a French-speaking West Indian who was required to submit to an oral interview and then denied a promotion because of his "peculiar" speech (*Loiseau vs. Dept. of Human Resources*, 1997. *Xieng vs. People's National Bank of Washington*, 1991 was related to a denial of a promotion based on the employee's heavy Cambodian accent.

Cases rendering opposite decisions in favor of employers include *Poskocil vs. Roanoke County School District*, 1999; *Fragante vs. Honolulu*, 1989; *Hou vs. Pa. Dept. of Education*, 1983; *Kureshy vs. City University of New York*, 1983; and *Tran vs. Houston*, 1983. A primary factor in the verdicts was the agencies' abilities to demonstrate links between language and jobs. Of particular interest to foodservice managers is a finding of no discrimination when, because of her poor spoken English, a Hispanic chambermaid was denied promotion to a cashier position in which she would have to interact with hotel guests (*Mejia vs. New York Sheraton Hotel*, 1979). At this point, we turn to other language-policy issues that appear in the workplace, specifically, those related to the use of English-only policies and the treatment of bilingual employees.

The English-Only Debate

Business operators must not set policies in a vacuum of knowledge about relevant historical events and public opinion. We believe that examining past and current attitudes and lobbying efforts related to English-only and bilingual stances will offer an appropriate backdrop from which to view the potential consequences of setting language policies in a foodservice organization. In order to comprehend the forces that may influence the actions of employers, employees, and other organizational actors in the foodservice business environment, we look at first the platforms of each side of the English-only debate occurring in the United States and secondly, the support for bilingual programs sponsored by public and private entities.

Proponents of the English-Only Movements

Many of the English-only supporters use the term "official language movement"; the campaign itself has been closely tied to anti-immigration, anti-bilingual education, and anti-affirmative action movements since the late 1970s. Three organizations lead the English-only movements in the United States: *U.S. English, English First,* and *ProEnglish.* Their individual visions or mission statements drawn from their websites follow:

- U.S. English, Inc. (2005) believes that the passage of English as the official language will help to expand opportunities for immigrants to learn and speak English, the single greatest empowering tool that immigrants must have to succeed.
- English First is a national, non-profit grassroots lobbying organization founded in 1986. Our goals are simple: make English America's official language, give every child the chance to learn English, and eliminate costly and ineffective multilingual policies.
- ProEnglish is the nation's leading advocate of official English. We work through the courts and in the court of public opinion to defend English's historic role as the common, unifying language of the United States of America, and to persuade lawmakers to adopt English as the official language at all levels of government.

Due to the close relationship between their attitudes toward immigration and those towards English as the official language for the United States, these organizations often ask 'why, when so many of the initial immigrants to the United States were capable of learning and working in English, does the public and private community have to continue to make significant financial investments in bilingual programs and materials for people who refuse to learn English and hinder the productivity of our nation?' Interestingly, "while polls have shown that American respondents seem to favor English as the official language for the United States, they are less apt to support restrictions on the use of minority languages or the termination of bilingual service to those who depend on them" (Crawford, J., 1996, March).

Across the board, much of these organizations' efforts have gone to lobby for laws declaring English the official language both at the federal and state level. Legislation prompted at the federal level has consistently been overturned so far, but "proponents have been successful at the state level. Beginning with Louisiana in 1812, 27 states have adopted some form of official-English law" (Dobbs, L., 2005, June 19). These state laws have had varied impacts on public services. In related actions, some states have eliminated bilingual education programs, e.g., California (Proposition 227) and Arizona (Proposition 203), but others continue to provide multilingual materials such as voter ballots, drivers' license tests, and other government publications due to legal loopholes built into their laws. The Iowa English Reaffirmation Act of 2001, for example, "exempts information relating to trade or tourism, public health and 'documents that protect the rights of victims of crimes or criminal defendants,' among other things and...[has] a general loophole allowing state employees to communicate in another language if it is necessary or desirable to do so" (Kelderman, E., 2003, November 14).

To date, the foodservice sector in the United States has not been unduly affected by these laws, unlike the case in Quebec "where French is to be given priority over English. For example, by law, French store signs must be twice the size of their English translations. And language laws are enforced in Quebec. Language police will issue citations to businesses breaking rules. This enforcement generates much anger particularly among small businesses that may not be able to follow the language laws all the time" (Maceri, 2005). It is also evident that, where customers are concerned, American marketers are steering in the opposite direction of English-only movements, following the siren call of the dollar from targeted non-English speaking minority consumers. Nevertheless, it has been suggested that if hard-core English-only supporters force businesses to comply with local ordinances mandating signs in English-only, foodservice operators could experience "restrictions on bilingual menus at fast food restaurants, French or Italian restaurants where the wait staff could only speak in English, or a city's Chinatown with signs in English-only" (National Conference for Community and Justice, 2002).

CRITICS OF THE "OFFICIAL LANGUAGE" MOVEMENTS

Organizations taking a position opposing English-only laws include but are not limited to the founders of the English Plus Information Clearinghouse: the American Civil Liberties Union, American Jewish Committee, American Jewish Congress, Caribbean Education and Legal Defense Fund, Center for Applied Linguistics, Chinese for Affirmative Action, Coloradans for Language Freedom, Committee for a Multilingual New York, Conference on College Composition and Communication, Christian Church (Disciples of Christ), El Concilio de El Paso, Haitian American Anti-Defamation League, Haitian Refugee Center, Image de Denver, IRATE (Coalition of Massachusetts Trade Unions), Mexican American Legal Defense and Educational Fund, Michigan English Plus Coalition, META (Multicultural Education, Training, and Advocacy) Inc., National Association for Bilingual Education, National Coalition of Advocates for Students, National Council of La Raza, National Puerto Rican Coalition, New York Association for New Americans, Organization of Chinese Americans, Spanish-Speaking/Surnamed Political Association, Stop English Only Committee of Hostos Community College, Teachers of English to Speakers of Other Languages (English Plus Information Clearinghouse, EPIC). The perspectives of members of these organizations vary; some comment wryly on the fact that:

> Americans use hundreds of words and phrases in our everyday speech that stem from other languages and other cultures. From our national motto of E Pluribus Unum to chili con carne, America's diversity is expressed millions of times each day through the colorful nature of the language of her people. Language is a common thread that can bind together the people of a country or can cause the national fabric of unity to unravel. (National Conference for Community and Justice, 2002)

Other critics assert that linguistic policy movements, such as U.S. English, are racist and treat non-English speakers as culturally and linguistically inferior. This is further underscored by Raul Yzaguirre, president of the National Council of La Raza: "US English is to Hispanics as the Ku Klux Klan is to blacks" (Hartman, 2002). Yzaguirre's comment gives us a keen understanding of the perspective that any laws or policies negating the right to use languages other than English in "official" settings such as public offices or business workplaces, have the potential to cause harm to minority or immigrant populations by continuing a legacy of suppression. Border towns on the U.S. side have long traditions of censuring Spanish in public sites, e.g., signs in foodservice establishments indicating 'Spanish not spoken here.' The case of a group of restaurant workers from a drive-in restaurant in Arizona who were asked to sign an agreement not to speak Navajo was particularly notorious since the ability to speak Navajo is rapidly disappearing among Native Americans. This fact was one of many that led to the Native American Languages Act of 1990 as a part of ongoing attempts to ameliorate past discrimination.

It is with observance of these competing philosophies towards English-only movements across the nation that we now turn to a discussion of employer-prescribed English-only language policies in the workplace. The tensions that are manifested between these competing movements are also surfacing in both public and private sectors where employees claim that English-only language policies are discriminatory under Title VII of the Civil Rights Code.

English-Only Policies in the Workplace

By far, the greatest number of clashes between employers and employees is on the subject of English-only policies in the workplace. However, there is divergent opinion between the EEOC and some of the legal rulings in the area of requiring English-only communication on the job. To understand this conflict better, we need to explore the underlying interests of both the employer and the employees. The business rationales for setting an English-only communication policy are several, including the business' desire to:

- improve employees' English proficiency to better serve a primarily English-speaking customer base;
- encourage workplace efficiency by guaranteeing an English-speaker supervisor's or coworker's ability to communicate with all employees,
- reduce ethnic tension, particularly where their staff has segregated itself along language lines or where customers or employees feel

non-English speakers are speaking badly of them or specifically harassing them;
- facilitate cooperative work assignments in which exclusive use of the English language is essential for efficiency; and
- ensure safety in the workforce.

Despite these prospective benefits, English-only policies can carry risks for the employer. EEOC guidelines state that, in certain situations, English-only rules may be viewed as discriminatory on the basis of national origin under Title VII of the 1964 Civil Rights Act. The EEOC suggests that English-only rules may actually increase ethnic tension in the workplace, rather than reduce it and has taken the position that English-only rules are harmful especially when they are applied at all times, including employee breaks and lunch periods. According to the EEOC, forbidding employees to speak the language they know best reduces the person's employment opportunities, and may create an "atmosphere of inferiority, isolation and intimidation."

However, the agency's guidelines also state that English-only rules are permissible when: (a) speaking a common language is imperative for safety and (b) it is a matter of business necessity, for example, if a person's lack of English skills would have a detrimental effect on job performance. We note here that the courts have not offered a consistent ruling for appraising the legitimacy of English-only policies. The legal outcome may depend on whether or not English-only policies are applied equally to all employees; the latter situation would be where Spanish-speaking staff members are disciplined for speaking Spanish, but workers who might speak German or Vietnamese are not bothered as in a case involving a nursing home, Royalwood Care Center. How a policy is implemented and whether there is an overall atmosphere of ethnic tension is also a critical factor in the court's decision. The EEOC backed several employee claims. One settled in 2003 for $ 1.5 million. The case involved Hispanic housekeepers at a casino who were not permitted to speak Spanish and a janitorial supervisor claiming he had to train Spanish-speaking hires in a closet to avoid language persecution. Others from the casino told of harassment by supervisors who called them "wetbacks," accused them of stealing, and fired them for objecting to the English policy. EEOC also filed against the owners of a group of Supercuts hair salons, due to factors including a sign warning that speaking in a language other than English is both disrespectful and prohibited, and against Sephora cosmetics stores, where employees could speak Spanish only with customers but not with each other, even on breaks.

At times decisions regarding English-only in the workplace evolve from management's response to customers or employees claiming that non-English speaking personnel are making rude comments about them. *Prado vs. L. Luria & Son, Inc.*, 1997 led to a determination that that the promotion of harmony in the workplace is a reasonable business justification for implementing an English-only rule. With respect to the rationale of customers preferring to hear only English in the business operation, *Jurado vs. Eleven-Fifty Corp*, 1987 rendered a verdict permitting a radio station to choose the language an announcer must use based on the fact that a majority of listeners preferred English. On the other hand, *Hernandez vs. Erlenbusch*, 1973 denied this defense, in which case the defendants, owners of a tavern, were known to have instructed their bartenders to restrict the use of non-English languages in the bar [ostensibly to avoid confrontations with] the regular [Anglo] trade and if there should be a chance of a problem, to ask the 'problem' people to move to a table and then turn the juke box up. The court equated this tavern's policy with the banishing of blacks to the back of the bus to avoid the racial animosity of dominant white passengers, given that it ordered Spanish-speaking patrons to the "back booth or out" to avoid provoking English-speaking beer-drinkers.

Even when safety is a reasonable defense, companies may find that the ramifications of setting English-only policies may be more disturbing to the organizational culture than realized initially. For example, Air France implemented a policy requiring its multilingual pilots to speak English exclusively when communicating with air traffic controllers at Charles de Gaulle Airport in Paris. Air France believed that this policy would improve communications and reduce the potential for airline accidents. Ultimately, employees criticized the policy saying that the English-only rule heightened communication problems between pilots and air traffic controllers because not all of Air France's employees were skilled in speaking English. In the wake of protests by pilots, air traffic controllers and their unions, Air France discontinued its English-only policy 15 days after its implementation (Daley, 2000). Furthermore, another court found that an English-only rule may have an adverse impact on employees when employers do not carefully explain the consequences of speaking languages other than English *(EEOC vs. Synchro-Start Products, Inc.*, (1999). Adverse impact occurs in a legal sense when a decision, practice, or policy has a disproportionately negative effect on a protected group, in this case on employees with no or extremely limited English-speaking skills.

Special Concerns Related to Bilingual Employees

At times, employees are hired because they speak a language other than English so that they may serve as translators for an operation's employees and clientele. For example, a restaurant with a majority of Spanish-speaking workers in the kitchen may want to have a supervisor or chef who is bilingual in English and Spanish. As seen previously, employers should be able to prove that the job in question actually requires fluency in the second language. Conflict can also arise with employees who are fired or demoted because they lack a second language that over the course of their tenure with an agency or firm became important to the position. Barry Johnson, a sanitation supervisor, claimed that he was demoted back to garbage truck detail (losing $1.52 an hour) so that the city could hire a Spanish-speaking individual in his stead due to an expanding number of non-English speaking clients in his geographic territory (Spielman, 2005).

Foodservice firms may hire and assign bilingual employees to communicate with non-English speaking customers or workers as long as their second language duties are clearly defined and there is evidence of corresponding compensation for this bilingual skill. However, when the bilingual employee is made to perform additional duties related to translation or interpretation without appropriate compensation to distinguish them from the non-bilingual employees in the same position, the courts have recognized disparate treatment. A police detective, Felipe Arroyo, was awarded $1.3 million due to a finding that excessive work was assigned to him because of his Spanish-speaking abilities (Moran, 2004). This case also clarifies that bilingual employees who are denied transfer or promotion opportunities or leave or vacation time because their bilingual skills are needed in the workplace can also claim disparate treatment. Their employers risk a claim of national origin discrimination where the employees declare that their bilingualism is a factor in denying them opportunities available to non-bilingual employees.

On a separate issue related to bilingualism, several courts have concluded that an English-only rule does not discriminate against bilingual employees because they can voluntarily comply with the rule. Examples of this are *Garcia vs. Spun Steak Co.* (1993), *Garcia vs. Gloor* (1980), *Long vs. First Union Corp.* (1995/1996), and *Kania vs. Archdiocese of Philadelphia,* (1998). Nonetheless, managers should note that the EEOC presented expert testimony in *EEOC vs. Premier Operator Servs. Inc.* (1999) regarding a linguistic phenomenon termed 'code-switching'; this testimony based on linguistic research asserted that bilingual persons, particularly those who grow up in bilingual environments, will tend to alternate between languages involuntarily. This unconscious return to one's native language can in certain situations be viewed as an immutable aspect of one's national origin. As seen in *EEOC vs. Premier Operator Servs. Inc.* excessive discipline actions, e.g., termination, based on slips of the employee's tongue will be examined critically by the EEOC. There is no doubt that treatment of bilingual employees is an area rife with implications for foodservice managers who do not take heed of lessons learned from these various court renderings and data from linguistic science.

CONCLUSION AND RECOMMENDATIONS TO FOODSERVICE OPERATORS

The primary lesson with respect to language policies in the foodservice workplace is that the foodservice manager must weigh the business rationale for a language policy against any discriminatory effects the policy might produce. Prior to setting a language policy, employers should visibly exhaust all possible alternatives that might accomplish the foodservice operator's goals. Businesses should document evidence to support any safety justification for their policies as well as evidence of any other business necessity for the policy. Operators should also avoid blanket policies that impact employees while they are on lunch or rest breaks or in non-working locations, e.g., changing clothes in the employee locker room.

In every instance, if the decision is made to institute an English-only policy in the workplace, management must clearly explain the business reasons for the policy to the employees. The employer must confirm that all employees are advised of the policy, the specific situations that it covers, and the penalties that will ensue for violating the policy. Employers are urged to use all available means of notification such formal meetings, e-mail, postings, notice with a paycheck or any other reasonable means that will reach one hundred percent of the staff. The EEOC also recommends that employers give this notice in both English and other native languages used by its employees and that a grace period be used before the rule is applied in order to make sure that all employees are aware of and understand the policy and its implications for them personally. Line managers will need to actively listen for negative responses related to the company's moral standing and concern for minority or immigrant workers.

As noted, the EEOC takes the position that employers must establish written objective criteria for reviewing applicant credentials during critical personnel decision processes. Companies need to apply the same criteria consistently across all candidates for hiring or promotion. Employment want ads should inform prospective candidates of all required qualifications, including any qualifications related to language ability, in addition to stating that the employer is an 'equal opportunity employer.' To adequately document the business necessity, management should identify all language tasks, e.g., reading, writing, and oral interaction, required to carry out the duties and responsibilities of the job. This determination is based on an objective assessment of the duties and responsibilities of the position. The list should specify both the language(s) and the proficiency levels in which each task must be accomplished. Before staffing a bilingual position, management should determine if the language requirements of the position must be met at the time of appointment or could be fulfilled with language training at a later date.

Additional precautions include: reviewing all personnel policies and informal practices related to language fluency and use; dropping all but essential language requirements; prohibiting discrimination against those with accents; advising superiors against discriminatory remarks; and distributing a written description of anti-discriminatory policies with a clear mention of language and national origin discrimination. For example, employers should avoid unfairly punishing bilingual employees for an accidental "slip of the tongue." We acknowledge that at times measures to manage diversity are responses to legal verdicts against companies. For example, Cracker Barrel, after paying hefty fines in court, now provides to its non English-speaking workers an interactive laptop which has English language modules they can study at home. On a positive note, however, a survey of 558 human resources professionals in the U.S. has found that 28% already offer language courses to their employees. An additional 27% of the sample firms have changed health and safety policies to respond to the language needs of the workforce, 27% are changing employment practices to address discrimination based on ethnicity, and 24% are offering language courses to managers (Schramm & Burke, 2004). There is definitely room for expansion of this activity among foodservice and other business sectors across the board.

In summary, noting the absence of consistent rulings from the courts, we have attempted to offer practical guidelines for foodservice operators. It is with this purpose that we have undertaken this review of the law and issues surrounding language policies in the U.S. workplace. Furthermore, since the foodservice industry is highly dependent on immigrant workers, our paper has given voice to the concerns and rights of these members of the American labor pool. At this point, having concluded the legal analysis which constituted the central purpose of this paper, we turn to a discussion of topics that merit further research efforts on the part of discipline experts in the field of hospitality management interested in organizational science and communication studies.

Propositions for Future Research

Although managers can make language policy decisions taking only into consideration the context of the legal framework outlined in this paper, they do so in the stark absence of information about the impacts of language policies on employees and other organizational actors. Moreover, a supplementary literature search in the broadest set of related disciplines including organizational behavior, human resources, immigration and minority studies, organizational communication, and cultural, language and translation studies has identified multiple relevant theoretical foundations for the study of language policy as a variable impacting organizational life in the business sector. To expand beyond the benefits of our current legal focus, we strongly recommend that researchers conduct future explanatory studies to determine how language policies may impact employees in foodservice organizations, particularly in the area of staff perceptions of both cultural climate and organizational management of diversity. Contemplating many of the unanswered questions arising during the courts' reflections on the English-only and bilingual employee matters, we offer several very generally stated propositions to spearhead the research in this direction. These propositions can be subdivided into more narrowly focused research topics as befits the interests and needs of a research team.

Proposition 1:

As linguistic policies in foodservice organizations move towards an English-only orientation, the following outcomes reflected in language minority employee behavior and perceptual attributions comprising (1) task accuracy and performance, (2) employee power, morale and stress levels and (3) perceptions of cultural climate and organizational responses to diversity will move towards their negative poles.

Proposition 2:

As linguistic policies in foodservice organizations move towards an English-only orientation, the following outcomes (1) *team performance* in which English is needed to promote efficiency, (2) *perceptions of cultural climate* of English-speaking supervisors and coworkers, and (3) English-speaking customers' *attributions* of language as a precursor of server helpfulness, politeness and approachability will move toward their positive poles.

Proposition 3:

Morale and stress levels of employees required to apply bilingual skills in the workplace will be correlated with (1) levels of added bilingual pay compensation, (2) employee language proficiency levels, (3) percentage of job or work hours involving the use of bilingual skills, and (4) perceived fairness of employee work and vacation schedules.

In addition, during the course of the current language policy research, a review of trade journals and newspapers revealed the following initial set of organizational strategies and communication interventions evident in foodservice establishments: formal or informal language policies; verbal translation strategies ranging from formal translation and interpretation to the informal use of interpreter banks or language brokers, e.g., a bilingual member of staff; poka yokes or fail-proofing devices such as color-coding chemical solvents for handling purposes; and random or acquired non-verbal gesturing tactics. A descriptive research project using interview or survey methods might readily be conducted across business or public organizations similar to foodservice structures to benchmark the various ways in which these organizations are successfully addressing their cross-lingual and multi-lingual communication requirements. An examination of organizational or human resources materials, e.g., employee manuals, posters, recruitment ads, would conceivably substantiate these interviews. We do recommend the use of screening questions in such a study to control for organizational size greater than 15 employees (subject to Title VII of the Civil Rights Act) and to ensure a study population where presence of linguistic diversity exists and where English as primary organizational language is the norm.

REFERENCES

Bahls, S. C. & Bahls, J. E. (1998). Watch your language if you require employees to speak only English, you'd better beware of the EEOC. *Entrepreneur Magazine*, December 1998. Retrieved October 5, 2005 from http://www.entrepreneur.com/article/0,4621,229602,00.html.

Berke vs. Ohio Dept. of Public Welfare, 52 Ohio App. 2d 271, 369 N.E.2d 1056, 1976 Ohio App., 6 Ohio Op. 3d 280 (10th District Ohio 1996).

Burns, W. C. (1995, 1996) Content Validity, Face Validity, and Quantitative Face Validity. Retrieved October 5, 2005 from http://www.burns.com/wcbcontval.htm.

Crawford, J. (1996, March). Anatomy of the English-only movement. Paper presented at a conference at the University of Illinois at Urbana-Champaign. Retrieved October 5, 2005 from http://ourworld.compuserve.com/homepages/JWCRAWFORD/anatomy.htm.

Daley, S. (2000). Roissy Journal; Pilots Just Say Non to English-Only. *NY Times*. May 23, 2000, Tuesday. Late Edition–Final, Section A, Page 4, Column 3. Retrieved October 5 from: http://select.nytimes.com/gst/abstract.html?res = F30611F7355-E0C708EDDAC0894D8404482.

Dobbs, L. (2005, June 19). English-only advocates see barriers to bill easing up. *CNN.com*. Retrieved October 5, 2005 from http://www.cnn.com/2005/US/04/18/official.english/.

E.E.O.C. vs. Premier Operator Servs. Inc., 75 F. Supp. 2d 550, 557 (N.D. Tex. 1999).

E.E.O.C. vs. Synchro-Start Products, Inc., 29 F. Supp. 2d 911, 913 (N.D. Ill. 1999).

English First. Retrieved October 5, 2005 from http://www.englishfirst.org/.

English Plus Information Clearinghouse (EPIC). The English Plus alternative. Retrieved October 5, 2005 from http://ourworld.compuserve.com/homepages/JWCRAWFORD/EPIC.htm.

Fragante vs. Honolulu, 494 U.S. 1081; 110 S. Ct. 1811; 108 L. Ed. 2d 942; 1990 U.S.

Frontera vs. Sindell, 522 F.2d 1215 (6th Cir. 1975).

Garcia vs. Gloor, 618 F.2d 264, 270 (5th Cir. 1980).

Garcia vs. Rush-Presbyterian St. Luke's Medical Center, 660 F.2d 1217 (7th Cir. 1981).

Garcia vs. Spun Steak Co., 998 F.2d 1480, 1487 (9th Cir. 1993).

Hartman, A. (2002). Language as oppression: The English-only movement in the United States. *Socialism and Democracy, 19*(2). Retrieved October 5, 2005 from http://www.sdonline.org/33/andrew_hartman.htm.

Hernandez vs. Erlenbusch, 368 F. Supp. 752; 1973 U.S. Dist.

Hou vs. Pa. Dept. of Education, 573 F. Supp. 1539, 1983 U.S. Dist., 33 Fair Empl. Prac. Cas. (W.D. PA 1983).

Jurado vs. Eleven-Fifty Corp, 813 F.2d 1406 (9th Cir. 1987).

Jackson, J. (2002). The restaurant industry, the largest employer of immigrants in the nation, hoping "Guest Worker Program" is a part of immigration reform. *Hotel Online: News for the Hospitality Executive*. Retrieved October 5, 2005 from http://instruct1.cit.cornell.edu/courses/ha191/Mar02_RestrEmployees.html.

Joyce, A. (2004). EEOC sues Virginia hotel over English fluency policy. Washington Post. Wednesday, October 6, 2004. Retrieved October 5, 2005 from: http://www.washingtonpost.com/wp-dyn/articles/A9663-2004Oct5.html.

Kania vs. Archdiocese of Philadelphia, 14 F. Supp. 2d 730, 733-34 (E.D. Pa. 1998).

Kelderman. E. (2003, November 14). Businesses outdo states in multi-lingual outreach. *Stateline.org*. Retrieved October 5, 2005 from http://www.stateline.org/live/ViewPage.action?siteNodeId = 136&languageId = 1&contentId = 15474.

Kureshy vs. City University of New York, 561 F. Supp. 1098, 1983 U.S. Dist., 31 Fair Empl. Prac. Cas. (E.D. NY 1983).

Little Forest Medical Center vs. Ohio Civil Right Commission, 61 Ohio St. 3d 607, 575 N.E.2d 1164, 1991 Ohio, 56 Fair Empl. Prac. Cas., 60 Empl. Prac. Dec (S.C. OH 1991).

Loiseau vs. Dept. of Human Resources, 1997 U.S. App. (9th Cir. Filed April 30, 1997).

Long vs. First Union Corp., 894 F. Supp. 933, 941 (E.D. Va. 1995), affirmed. 86 F.3d 1151 (4th Cir. 1996).

Lydecker, T. (1996). Communicating in a melting pot. *Restaurants USA*. Retrieved October 5, 2005 from http://www.restaurant.org/business/magarticle.cfm?ArticleID = 106.

Maceri, D. (2005). Spanish. *The Seoul Times*. Retrieved October 5, 2005 from http://theseoultimes.com/ST/?url = /ST/db/read.php?idx = 189.

Mejia vs. New York Sheraton Hotel, 449 U.S. 854; 101 S. Ct. 149; 66 L. Ed. 2d 67; 1980 U.S.

Moran, G. (2004) Spanish skills led to extra work, S.D. detective says. Signs on San Diego. May 13, 2004. Retrieved October 5, 2005 from: http://www.signonsandiego.com/uniontrib/20040513/news_7m13discrim.html.

National Conference for Community and Justice (2002). English-only: What are the issues? Retrieved October 5, 2005 from http://www.nccj.org/nccj/nccj.nsf/subarticleall/393?opendocument.

Poskocil vs. Roanoke County Sch. Div., Civil Action No. 98-0216-R, United States District Court for the Western District of Virginia, Roanoke Division. 1999 U.S. Dist.

Prado vs. L. Luria & Son, 975 F. Supp. 1349; 1997 U.S. Dist.

ProEnglish: English Language Advocates. Retrieved October 5, 2005 from http://www.proenglish.org/.

Schramm J. & Burke, M. E. (2004) Workplace Forecast: A Strategic Outlook. *Research report published by the Society for Human Resources Management*. June 2004.

Shin, H. B. & Bruno, R. (2003). Language use and speaking ability: 2000: Census 2000 brief. Retrieved October 5, 2005 from http://www.census.gov/prod/2003pubs/c2kbr-29.pdf#search = 'US%20census%20data%20and%20nonenglish%20speakers.'

Smith, Gerrit B. (2005). Note. I want to speak like a native speaker: the case for lowering the plaintiff's burden of proof in Title VII accent discrimination cases. 66 Ohio St. L.J. 231-267.

Spielman, F. (2005). Streets & San worker calls 12th Ward demotion garbage Chicago Sun Times. August 19, 2005. Retrieved October 5 from: http://www.suntimes.com/output/news/cst-nws-demote19.html.

Stern, M. (2003). INS jumps estimate of illegal immigrants: Number put at 8 million; growth rate is increased. *Sign on San Diego.com*. Retrieved October 5, 2005 from http://www.signonsandiego.com/news/mexico/20030131-9999_1n31immig.html.

Tran vs. Houston, 1983 U.S. Dist., 35 Fair Empl. Prac. Cas., 31 Empl. Prac. Dec. (S.D. TX, Houston Div.1983).

U.S. English, Inc. (2005). About U.S. English. Retrieved October 5, 2005 from http://www.us-english.org/inc/about/.
U.S. Equal Employment Opportunity Commission (2002). Section 13: National origin discrimination. *EEOC compliance Manual*. Retrieved October 5, 2005 from http://www.eeoc.gov/policy/docs/national-origin.html.
Vasquez vs. McAllen Bag & Supply Co., 660 F.2d 686 (5th Cir. 1981).
Xieng vs. Peoples Nat'l Bank, 119 Wn.2d 1001; 832 P.2d 488; 1992 Wash.

doi:10.1300/J369v09n02_03

The Role of Language Fluency Self-Efficacy in Organizational Commitment and Perceived Organizational Support

Amy Van Dyk
Priscilla Chaffe-Stengel
Rudolph J. Sanchez
Julie B. Olson-Buchanan

SUMMARY. This study explored the relationship between language fluency self-efficacy and organizational commitment and perceived or-

---

Amy Van Dyk is a graduate of the Craig School of Business at California State University, Fresno.

Priscilla Chaffe-Stengel, PhD, is Stanford University Professor, Information Systems and Decision Sciences Faculty Fellow, Institutional Research, Assessment and Planning, California State University, Fresno (Email: pchaffe@csufresno.edu).

Rudolph J. Sanchez, PhD, is Assistant Professor of Management, Craig School of Business, California State University, Fresno, 5245 N. Backer Ave. M/S PB7, Fresno, CA 93740 (E-mail: rjsanchez@csufresno.edu).

Dr. Julie B. Olson-Buchanan earned a PhD in Industrial-Organizational Psychology from the University of Illinois, Urbana-Champaign. She is currently Department Chair and Professor in the Department of Management, in the Craig School of Business at California State University, Fresno. She can be reached by email at julieo@csufresno.edu.

All correspondence regarding the article should be directed to Rudolph J. Sanchez at the above address.

[Haworth co-indexing entry note]: "The Role of Language Fluency Self-Efficacy in Organizational Commitment and Perceived Organizational Support." Van Dyk, et al. Co-published simultaneously in *Journal of Foodservice Business Research* (The Haworth Hospitality & Tourism Press, an imprint of The Haworth Press, Inc.) Vol 9, No. 2/3, 2006, pp. 49-66; and: *Human Resources in the Foodservice Industry: Organizational Behavior Management Approaches* (ed: Dennis Reynolds, and Karthik Namasivayam) The Haworth Hospitality & Tourism Press, an imprint of The Haworth Press, 2006, pp. 49-66. Single or multiple copies of this article are available for a fee from The Haworth Document Delivery Service [1-800-HAWORTH, 9:00 a.m. 5:00 p.m. (EST). E-mail address: docdelivery@haworthpress.com].

Available online at http://jfbr.haworthpress.com

doi:10.1300/J369v09n02_04

ganizational support in a sample of non-native English speaking employees doing food preparation work. A reliable, single-factor measure of language fluency self-efficacy was developed for this study. Consistent with existing literature, we found a significant relationship between language fluency self-efficacy and perceived organizational support. Additionally, perceived organizational support was related to organizational commitment. Structural equation modeling analysis results showed that the data were a good fit to the model. doi:10.1300/J369v09n02_04 *[Article copies available for a fee from The Haworth Document Delivery Service: 1-800-HAWORTH. E-mail address: <docdelivery@haworthpress.com> Website: <http://www.HaworthPress.com>*

KEYWORDS. Language fluency self-efficacy, organizational commitment, perceived organizational support

INTRODUCTION

Several researchers have noted the changing demographics of the United States workforce, particularly with respect to such factors as age and ethnicity. Indeed, a substantial amount of research has focused on how these changing demographics have potential implications for the continued effectiveness of our best practices in management (e.g., Lee-Ross, 2005; Stone-Romero, Stone, & Salas, 2003). Yet, very little attention has been paid to how management practices may be affected by the increasing proportion of non-native English speakers in the U.S. workforce. For example, according to the 2000 United States Census, there are 4.3 million Spanish-only speaking people (over 13% of the population) in the State of California alone. Clearly, this population is an increasingly important talent resource for business generally and the foodservice industry specifically. Despite the increasing importance of this talent pool, few researchers have explored the role of language in traditional human resource and organizational behavior phenomena. Because the development of key employee attitudes is theoretically dependent on communication between the organization and its representatives and employees, the ability of employees to understand these communication messages is critical. In organizations where multiple languages are spoken, this assumption needs to be explored.

The present study applies two well-established organizational behavior theories with a sample of non-native English speaking employees doing food preparation work. Using the theoretical foundations of self-efficacy

(Bandura, 1977; 1997) and social exchange (Blau, 1964), this study examines the relationship between the participant's English fluency self-efficacy and perceived organizational support (POS) (Eisenberger, Huntington, Hutchison, & Sowa, 1986) and organizational commitment (OC) (Meyer, Allen, & Smith, 1993). We focused on these particular constructs because prior research from a variety of industries, has demonstrated that perceived organizational support and organizational commitment are important antecedents of several organizational and employee outcomes including job satisfaction, performance, and turnover intentions (e.g., Meyer, Stanley, Herscovitch, & Topolyntsky, 2002; Rhoades & Eisenberger, 2002). In the following sections, we will present the theoretical framework of self-efficacy; link language fluency self-efficacy to the development of perceived organizational support and organizational commitment; and finally, link perceived organizational support to the development of organizational commitment.

LANGUAGE FLUENCY SELF-EFFICACY

Self-efficacy is defined as an individual's subjective belief that he or she can complete a given task. Importantly, Bandura (1997) asserts that self-efficacy beliefs are activity specific. In the work setting, it is possible that an employee has high self-efficacy regarding job performance, but not in speaking with colleagues and organizational superiors. Although people may assume the primary predictor of performance on a task is a person's task ability, Bandura's work demonstrates that the extent to which an individual believes he or she can complete a given task (self-efficacy) is also an important predictor of task performance. This occurs because self-efficacy governs the way individuals pursue their goals, and with how much persistence they do so, as well as the effort they put forth in common tasks (O'Neil & Mone, 1998). The relationship between task ability and self-efficacy was explored by Collins (1982) with a sample of children regarding math performance. He found that in addition to ability, perceptions of self-efficacy were positively related to performance. Additionally, in a meta-analysis, Chen, Casper, and Cortina (2001) found that self-efficacy mediated the well-documented cognitive ability-performance relationship for relatively simple tasks. Reynolds (2002) found a causal link between foodservice managers' work-related self-efficacy and performance. Clearly, self-efficacy is an important factor in predicting and understanding task performance.

In this context, the "task performance" we are primarily interested in is language fluency or how well non-native English speaking employees communicate with their native English speaking supervisors and top administrators. To that end, we focus specifically on employees' language fluency self-efficacy.

One of the primary reasons we focus on language fluency self-efficacy in this study is because a central part of Bandura's well-supported social cognitive theory asserts that self-efficacy can predict and explain many human behaviors (Bandura, 1986). For example, in the work environment, Judge and colleagues (Judge & Bono, 2001; Judge, Locke, & Durham, 1997) have shown that self-efficacy is related to job satisfaction and job performance. Theoretically, because employees who have high task specific self-efficacy are likely to persist in the face of obstacles, they are more likely to have success in that task (e.g., their job) which leads to increased satisfaction (Gist & Mitchell, 1992). Utilizing this same theoretical logic, we expect that language self-efficacy will be related to attitudes toward the organization such as OC and POS.

PERCEIVED ORGANIZATIONAL SUPPORT (POS)

POS has a rich theoretical background drawing from the work of Blau (1964) and Levinson (1965). POS refers to an employee's belief that he or she is valued by the organization (Eisenberger et al., 1986). Levinson asserted that individuals in work organizations attribute (or generalize) the actions of organizational representatives (e.g., supervisors and managers) to the organization. This "personification" of the organization psychologically prepares individuals to engage the organization in ways that mirror human social exchange relationships. Based on the early work of Levnison, Eisenberger et al. (1986) and his colleagues argued that employees draw from their interactions with organizational representatives to develop "global beliefs" about the extent to which their organization values them as workers as well as human beings.

In order for employees to perceive organizational support, it is necessary for them to understand what the organization is doing to provide support. Organizations engage in a variety of tactics to communicate support of employees including many written and verbal messages. We argue that the relationship between the organization and the employee can be hindered if an employee demonstrates a low level of self-efficacy concerning language fluency in the primary language of the business (i.e., the language spoken by top management). In a review of the POS literature, Rhoades and Eisenberger (2002) assert that the antecedents of POS include fair treatment of the employee, supervisor support, and organizational rewards and job conditions. To the extent that organizations communicate fair treatment via verbal and written messages (versus behavior), it becomes more difficult for employees to perceive fair treatment and consequently develop perceptions of organizational support.

Bhanthumnavin (2003) found that POS, task-specific self-efficacy, and location of workplace were correlated with higher performance (i.e., performance ratings and work effectiveness) in Thai work units. Additionally, self-efficacy was related to POS. Although not the focus of Bhanthumnavin's work, this finding further supports our assertion regarding the relationship between self-efficacy and POS.

> *Hypothesis 1*: Language fluency self-efficacy will be positively related to POS.

ORGANIZATIONAL COMMITMENT

Organizational commitment (OC) refers to the degree to which an employee feels a sense of loyalty to an organization (Allen & Meyer, 1990). The literature indicates that OC is related to job performance (e.g., Bauer & Green, 1998), turnover (e.g., Cohen, 1993), and job satisfaction (Shore & Martin, 1989). Sneed and Herman (1990) examined the relationship between job characteristics, organizational commitment, job satisfaction, and some demographic variables among supervisory and nonsupervisory employees in the hospital food service industry. They found a positive relationship between organizational commitment and job satisfaction, which supports the relationship found in other research involving these two constructs (e.g., Cheng & Stockdale, 2003; Camp, 1993). Clearly, given its relation to other important outcome variables (e.g., performance, turnover), organizations would benefit from understanding the antecedents of OC or understanding how OC might be fostered in the workplace.

Similar to the development of POS, the development of OC is partly dependent of employees properly understanding communication messages sent by the organization. For example, one of the components of OC is value congruence between the organization and employees (Mowday, Steers, & Porter, 1979). Most organizations will likely deliver a number of messages to employees to communicate organizational values (e.g., vision and mission statements) and to advise employees of how loyalty (i.e., organizational commitment) is valued and rewarded. If employees have low self-efficacy regarding language fluency in the language that these organizational messages are delivered, it is likely that the development of OC will be hindered. Therefore, we hypothesize:

Hypothesis 2: Language fluency self-efficacy will be positively related to OC.

Perceived Organizational Support and Organizational Commitment

There is both theoretical and empirical support for the assertion that POS is related to OC. From a theoretical standpoint, OC and POS operate within the larger umbrella of social exchange theory (e.g., Blau, 1964; Eisenberger, Stinglhamber, Vandenberghe, & Sucharshi, 2002). Social exchange theory states that when an individual feels support coming from the organization, he or she will feel obligated to that organization and develop commitment to the organization in return for its support. If the organization exhibits supportive behavior (i.e., human resource practices, rewards, procedural justice, etc.), the employee will report perceived support. This perceived support in turn leads to commitment to the organization (Eisenberger, Fasolo, & Davis-LaMastro, 1990; Shore & Wayne, 1993). That is, due to the norm of reciprocity, employees who feel they are supported by the organization are more likely to feel attached to the organization (Rhoades & Eisenberger, 2002) and subsequently develop attitudes and engage in behaviors that benefit the organization as a means of reciprocating.

Empirical evidence provides further support for the relation between POS and OC as predicted by social exchange theory. Social exchange theory has received support from a number of studies including one that collected data from manufacturing employees who exhibited a positive relationship between POS, commitment, performance, and constructiveness of suggestions made to the organization (Eisenberger et al., 1990). Similarly, in 2004, Sherman found that there were "consistently strong relationships between job involvement, POS and OC" in both full-time and part-time employees. (Sherman, 2004). Indeed, there is a consistent positive relationship between OC and POS throughout the literature

(Rhoades, Eisenberger, & Armeli, 2001; Settoon, Bennett, & Liden, 1996).

Other research suggests POS and OC are similarly related to other variables. Liao, Joshi and Chuang (2004) studied the relationship of dissimilarities in ethnicity, agreeableness and openness to experience in conjunction with OC and POS as well as other variables. They found that ethnic dissimilarity was negatively related to both OC and POS, dissimilarities in agreeableness negatively predicted POS, dissimilarities in openness to experience negatively predicted both OC and POS (Laio, et al., 2004).

Based on Bandura's theoretical work (1986), social exchange theory, and prior empirical work (e.g., Allen & Bradly, 2003; Bogler & Somech, 2004; Varona, 1996), we hypothesize:

Hypothesis 3: POS will be positively related to OC.

Model Development

In addition to testing the individual relationships between language self-efficacy and POS, and POS and OC, we tested a model that included all of the hypothesized relationships between language self-efficacy and POS and OC, and then POS to OC. Allen and Bradly studied OC and POS in organizations with increased communication due to the implementation of a TQM program versus organizations that did not have increased communication, and found that both OC and POS were higher in the organizations with increased communication. When the communication variable was examined in relation to OC and POS, the researchers found that a significant relationship existed. These findings give some basis to suspect that language fluency might similarly influence levels of OC and POS within an organization, because language fluency would affect communication (Allen and Bradly, 1997).

Methodology

A survey was distributed to approximately 250 food preparation workers a large frozen food manufacturer. The workers are primarily responsible for assembling food products (e.g., wrapped sandwiches). Thus, the tasks (and the corresponding skills) required to complete this work are similar to those in other food preparation areas in the foodservice industry (e.g., restaurants, catering, small markets). Participants were informed that their participation was strictly voluntary and separate from the employer. 112 surveys were returned, of which 109 were usable (response rate = 43.6%). Surveys in both Spanish and English were provided to participants so that they could respond to the survey in their preferred language. Most of the workers who participated in the survey (76%) spoke Spanish as their primary language. This was the sample used for all analyses (n = 89). Of the 88 respondents who returned usable surveys, 34 were male, 45 were female and 9 declined to answer the question. The mean length of service to the organization was 5.46 years, with a range of 19 years.

Measures

Language Fluency Self-efficacy. Because no measure of language self-efficacy in the work setting could be found in the literature, we developed a set of items that were theoretically consistent with Bandura's concept of self-efficacy within the context of language skills in the workplace. The language fluency self-efficacy measure was developed and tested using criteria from the Guidelines for Student Oral Language Observation Matrix (Alexandrowicz, 2004). *Language fluency self-efficacy*

TABLE 1. Factor Analytic Results for Language Fluency Scale

Item	**Factor 1 Language Fluency Self-efficacy**
When conversing in English, I have no trouble understanding what is being said.	.89
When conversing in English, I have no trouble being understood.	.88
I speak English very well. Thinking of the right words is effortless.	.87
When conversing in English with my supervisors, I think they have no difficulty understanding me.	.84
I feel that I speak as well as a native speaker.	.82
My social conversations are equally fluent in English and my native language.	.67
I consider myself equally fluent in English and my native language.	.63
Most of my social conversations at work are in English.	.38
Eigenvalue	4.74
Variance Accounted for	47.38

refers to the participant's belief that he or she can understand what is being said and can be understood in conversation.

Because we theoretically wanted to assess a unidimensional construct, principal components factor analysis was performed with a single factor forced on 10 items. Table 1 shows the results of the factor analysis. Factor loadings greater than .30 are reported. Reliability analysis revealed that the single factor scale had a reliability coefficient of .89 and it accounted for 47.38% of the variance in the items. An index for scale was created by taking the mean of all the responses to the items.

Organizational Commitment. Twelve items were taken from the Organizational Commitment Questionnaire (Mowday et al., 1979) to measure this construct ($\alpha = 0.83$). A sample item is "I find that my values and the organization's values are very similar." An index for scale was created by taking the mean of all the responses to the items.

Perceived Organizational Support. Six items from the Survey of Perceived Organizational Support (Eisenberger et al., 1986) were utilized to measure POS ($\alpha = 0.90$). A sample item is "The organization really cares about my well-being." An index for scale was created by taking the mean of all the responses to the items.

Once the complete survey was compiled in English, it was translated into Spanish in its entirety along with the cover letter. It was then back-translated to ensure accuracy.

Control Variables

In addition to the constructs of primary interest, we collected data on several variables that may influence the relationships between language efficacy and the attitudinal outcomes studied here. These variables include demographics (i.e., gender, tenure with the organization) and whether or not the employee believed that his or her organizational superiors' (supervisor and manager) native language was the same as his or hers.

RESULTS

The means, standard deviations, and intercorrelations for all study variables are reported in Table 2.

TABLE 2. Descriptive Statistics and Intercorrelations Between Study Variables

	Mean	SD	1	2	3	4	5	6	7
1 Gender	0.43	0.50	—						
2 Organizational Tenure	5.47	5.17	.03	—					
3 Supervisor Language Match	0.94	0.24	–.30	.00	—				
4 Manager Language Match	0.60	0.49	–.26	–.08	.21	—			
5 Language Fluency Self-Efficacy	3.69	1.18	–.13	.09	–.14	.10	(.89)		
6 Organizational Commitment	4.34	1.56	–.20	–.25	.13	.20	.23	(.83)	
7 Perceived Organizational Support	4.60	1.17	–.13	–.05	.01	.05	.45	.64	(.90)

Note: $N = 88$. Gender is coded Women = 0, Men = 1. Supervisor and Manager Language Match is coded 0 = superior does not have the same native language as employee, 1 = superior does have the same native language as employee. For the purpose of this study, the term supervisor represents the person a worker reports to directly, while the term manager refers to top management in the company. Numbers along the diagonal are scale reliabilities. Correlations >.19 are statistically significant at $p < .05$, correlations >.24 are significant at $p < .01$.

TABLE 3. The Relationship Between Language Fluency Self-Efficacy and Perceived Organizational Support

	B	R^2	ΔR^2	$F\Delta$
Step 1		.10		1.86
Gender	–.05			
Organizational Tenure	–.23*			
Supervisor Language Match	.13			
Manager Language Match	.14			
Step 2		.16	.06	4.92*
Language Fluency Self-efficacy	.26*			

Note: $N = 88$. Gender is coded women = 0, men = 1.
*$p < .05$.
**$p < .01$.

TABLE 4. The Relationship Between Language Fluency Self-Efficacy and Organizational Commitment

	B	R^2	ΔR^2	$F\Delta$
Step 1		.02		.83
Gender	–.13			
Organizational Tenure	.09			
Supervisor Language Match	–.01			
Manager Language Match	–.01			
Step 2		.22	.24	19.56**
Language Fluency Self-efficacy	.50**			

Note: $N = 88$. Gender is coded women = 0, men = 1.
**$p < .01$.

Language Fluency Self-Efficacy and Perceived Organizational Support

After controlling for gender, organizational tenure, and perceived language match, hierarchical regression analysis indicated that language fluency self-efficacy was related to perceived organizational support ($\Delta R^2 =$ .06, F = 4.91, $p < .05$). These results provide support for Hypothesis 1 (see Table 3).

TABLE 5. The Relationship Between Perceived Organizational Support and Organizational Commitment

	B	R^2	ΔR^2	$F\Delta$
Step 1		.02		.37
Gender	–.04			
Organizational Tenure	.18			
Supervisor Language Match	–.06			
Manager Language Match	–.09			
Step 2		.39	.37	41.37**
Language Fluency Self-efficacy	.64**			

Note: $N = 88$. Gender is coded women = 0, men = 1.
**$p < .01$.

FIGURE 1. Hypothesized Model Tested Using Structural Equation Modeling with Path Coefficients

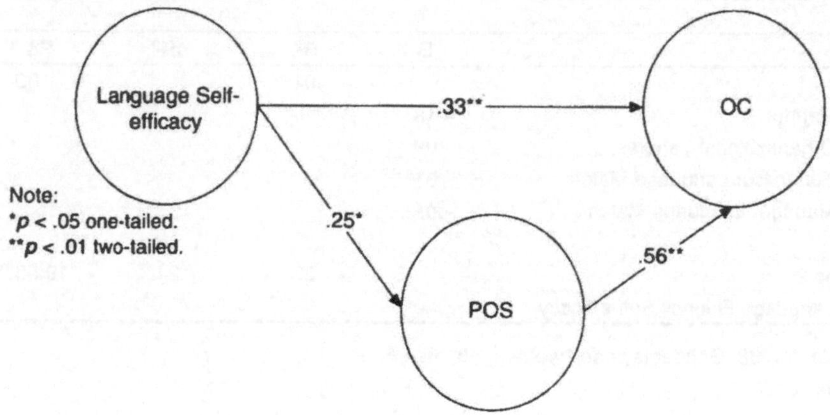

Language Fluency Self-Efficacy and Organizational Commitment

After controlling for gender, organizational tenure, and perceived language match, hierarchical regression analysis indicated that language fluency self-efficacy was related to organizational commitment ($\Delta R^2 = .22$, $F = 19.56$, $p < .01$). These results provide support for Hypothesis 2.

POS and OC

After controlling for gender, organizational tenure, and perceived language match, hierarchical regression analysis indicated that perceived organizational support was related to organizational commitment ($\Delta R^2 = .37$, $F = 41.38$, $p < .01$). This result provides support for Hypothesis 3.

Full Model Results

The model presented in Figure 1 was tested using structural equation modeling. Because we were interested in testing the structural paths in the model and not the more complex measurement model, we used single indicators (scale means) for the language self-efficacy and OC constructs. This has the effect of increasing the subjects to degrees-of-freedom ratio to provide better power to the test the hypothesized paths and full model. Consistent with current practice, the path from the indicator to the latent variable was set to the square root of the scale reliability. Additionally, the error variance was set to variance of scale multiplied by one minus scale reliability (Hayduk, 1987; Jorskog & Sorbom, 1989).

The chi-square statistic of the hypothesized model was statistically significant ($X^2 = 31.02$, $p < .01$) indicating a poor fit of the data. However, other indices of model fit indicated that the data fit the model well (GFI = .90; CFI = .93). Additionally, each of the structural paths was significant.

DISCUSSION

The purpose of this study was to explore the relationship between self-efficacy language fluency and two important attitudinal outcomes: organizational commitment and perceived organizational support. To that end, a reliable, single-factor scale of language fluency self-efficacy was developed for this study. Consistent with our first hypothesis drawn from Bandura's work (1977; 1986) and the POS literature (Eisenberger et al., 1986), we found a significant relationship between language fluency self-efficacy and perceived organizational support. When an employee has less confidence that he or she can communicate well in the primary language in which business is conducted, it is likely that the employees will receive less information from the organizational representatives that would convey the extent to which the organization values the employees as a worker as well as a human being. All of the participants in this study worked in the same organization, presumably with similar levels of supportive behavior from the organization (i.e., human resource practices, rewards, procedural justice, etc.). Yet employees with less language fluency self-efficacy reported lower POS, underscoring the importance of language in perceptions of POS.

Similarly, consistent with our second hypothesis developed from the OC (Mowday, et al., 1979) and self-efficacy literature (Bandura, 1977; 1986), language fluency self-efficacy was significantly related to OC. Employees with less confidence about their ability to communicate in English reported less commitment to the organization. This suggests that a language barrier between employees and organizational representatives may serve to inhibit the development of OC. Given the relation of OC to such factors as job performance and turnover, a language barrier may, indeed, be costly.

Finally, consistent with our third hypothesis drawn from both the POS and OC literature, the level of reported POS was significantly related to the level of reported OC in a sample of non-native English speakers working in an organization that conducts its business in English. This finding is consistent with the social exchange literature (e.g., Blau, 1964; Eisenberger et al., 2002) that asserts individuals who feel support coming from an organization will reciprocate this support by feeling committed to it (Eisenberger et al., 1990; Shore & Wayne, 1993).

The test of a model including the three constructs of interest allows for the simultaneous examination of the hypothesized relationships. Based on the results, we conclude that the hypothesized constellation of relationships explains the data collected. Interestingly, the relationship between language self-efficacy and OC was stronger that the relationship between language self-efficacy and POS. Future research can expand the number and type of constructs examined here. For example, there is strong evidence that an employee's relationship with his or her supervisor (e.g., leader-member exchange relationship; LMX) is important to attitudes and behavior in the workplace. It would be interesting to examine the role of language fluency self-efficacy on LMX.

IMPLICATIONS

The present research underscores the importance of communication in the workplace in relation to key POS and OC. Previous research in this area is generally based on organizations that have a common language, or assume a common language. As the workforce population changes and the primary languages of employees (and organizations) vary, the issue of language fluency becomes increasingly important. This study highlights that language fluency is not only important in terms of communicating essential organizational information (e.g., safety, job training), but may also be an important correlate of attitudes toward the organization. Because these attitudes have been linked with financial outcomes for organizations (e.g., productivity, turnover), organizational leaders working in multilingual environments should consider the broader implications of language fluency self-efficacy. This may mean that organizations consider ways to bridge gaps in fluency, thereby maximizing the gains of providing a supportive environment for employees.

Although this research does not imply a specific direction for human resource policies regarding language, it does suggest that organizations need to carefully examine language policies. Future research might examine the role of specific policies (e.g., only one language permitted/encouraged in the workplace; non-native speakers of primary language encouraged to utilize the language they are comfortable with) in relation to attitudinal and behavioral outcomes. This research also suggests that organizations consider human resource policies that focus on language fluency, especially self-efficacy, in order to increase OC and POS.

LIMITATIONS

The focus of this study was on native Spanish-speaking employees. Clearly, the theoretical assertions made here could be tested in other non-native English speaking groups. As an initial foray into the role of language fluency in attitudes toward the organization, we focused on a small set of constructs. We encourage future research that includes additional outcome variables.

CONCLUSION

This study suggests that language barriers in the workplace require special measures to ensure that employees feel connected to the organization in the forms of commitment and perceived support. The type of information provided by this study can help to pinpoint problems management may be facing, particularly high employee turnover which can lead to cost savings.

REFERENCES

Allen, N., & Meyer, J. (1990). The measurement and antecedents of affective continuance and normative commitment to the organization. *Journal of Occupational Psychology*. United Kingdom: British Psychological Society.

Bandura, A. (1997). *Self-efficacy: The Exercise of Control*. New York: Freeman and Company.

Bandura, A. (1998). Personal and collective efficacy in human education and change. *Advances in Psychological Science*. Col. 1: Social Personal and Cultural Aspects: 51-71. Hove, England: Psychology/Erlbaum (UK) Taylor and Francis.

Bandura, A. (1986). The explanatory and predictive scope of self-efficacy theory. *Journal of Social and Clinical Psychology, 4*, 359-373.

Bauer, T. N., & Green, S. G. (1998). Testing the combined effects of newcomer information seeking and manager behavior on socialization. *Journal of Applied Psychology, 83*, 72-83.

Bhanthumnavin, D. (2003). Perceived social support from supervisors and group members psychological and situational characterizes as predictor of subordinate performance in thai work units. *Human Resource Development Quarterly, 14*, 79-97.

Blau, P. M. (1964). *Exchange and Power in Social Life*. New York: Wiley.

Castillo, E. M. (2003). Psychosocioculural predictors of academic persistence decisions for Latino adolescents. Dissertations-Abstracts-International: Section B: *The Sciences and Engineering, 63*, (8-B) 3898.

Camp, S. D. (1993). Assessing the effects of organizational commitment and job satisfaction on turnover: an event history approach. *The Prison Journal, 74*, 279-305.

Chen, G., Casper, W. J., & Cortina, J. M. (2001). The roles of self-efficacy and task complexity in the relationships among cognitive ability, conscientiousness, and work-related performance: A meta-analytic examination. *Human Performance, 14*, 209-230.

Cheng, Y., & Stockdale, M. S. (2003). The validity of the three component model of organizational commitment in a Chinese context. *Journal of Vocational Behavior, 62*, 465-489.

Cohen, A. (1993). Organizational commitment and turnover: A meta-analysis. *Academy of Management Journal, 36*, 1140-1157.

Collins, J. L. (1982, March). *Self-efficacy and ability in achievement behavior*. Paper presented at the annual meeting of the American Educational Research Association, New York.

Eisenberger, R., Huntington, R., Hutchison, S., & Sowa, D. (1986). Perceived organizational support. *Journal of Applied Psychology, 7*, 500-507.

Eisenberger, R., Fasolo, P., Davis, P., & Davis-LaMastro, V. (1990). Perceived Organizational Support and Employee Diligence, Commitment and Innovation. *Journal of Applied Psychology, 75*, 51-59.

Eisenberger, R., Stinglhamber, F., Vandenberghe, C., Sucharski, I. L., & Rhoades, L. (2002). Perceived supervisor support: contributions to perceived organizational support and employee retention. *Journal of Applied Psychology, 87*, 565-573.

Gist, M. E., & Mitchell, T. R. (1992). Self-efficacy: A theoretical analysis of its determinant and malleability. *Academy of Management Review, 17*, 183-211.

Hayduk, L. A. (1987). *Structural equation modeling with LISREL.* Baltimore: Johns Hopkins University Press.

Jöreskog, K. G., & Sörbom, D. (1989). *LISREL 7: A guide to the program and applications.* Chicago: SPSS.

Judge, T. A., & Bono, J. E. (2001). The relationship between core self-evaluations traits–self-esteem, generalized self-efficacy, locus of control, and emotional stability–with job satisfaction and job performance: a meta-analysis. *Journal of Applied Psychology, 86*, 80-92.

Judge, T. A., Locke, E. A., & Durham, C. C. (1997). The dispositional causes of job satisfaction: A core evaluations approach. *Research in Organizational Behavior, 19*, 151-188.

Lee-Ross, D. (2005). Perceived job characteristics and internal work motivation: An exploratory cross-cultural analysis of the motivational antecedents of hotel workers in Mauritius and Australia. *Journal of Management Development, 24*, 253-266.

Levinson, H. (1965). Reciprocation: The relationship between man and organization. *Administrative Science Quarterly, 9*, 370-390.

Liao, H., Joshi, A., & Chuang, A. (2004). Sticking out like a sore thumb: employee dissimilarity and deviance at work. *Personnel Psychology, 57*, 969-1000.

Meyer, J. P., Allen, N. J., & Smith, C. A. (1993). Commitment to organizations and occupations: extension and test of a three-component conceptualization. *Journal of Applied Psychology*. June 538-551.

Mowday, R. T., Steers, R. M., & Porter, L. W. (1979). The measurement of organizational commitment. *Journal of Vocational Behavior, 14*, 224-227.

O'Neill, B. S., & Mone, M. A. (1998). Investigation of equity sensitivity as a moderator of relations between self-efficacy and workplace attitudes. *Journal of Applied Psychology, 83*, 805-816.

Reynolds, D. (2002). The moderating effect of leader-member exchange in the relationship between self-efficacy and performance. *Journal of Human Resources in Hospitality & Tourism, 3*, 77-90.

Rhoades, L., & Eisenberger, R. (2002). Perceived organizational support: A review of the literature. *Journal of Applied Psychology, 87*, 698-714.

Rhoades, L., Eisenberger, R., & Armeli, S. (2001). Affective commitment to the organization: the contribution of perceived organizational support. *Journal of Applied Psychology, 86*, 825-836.

Settoon, R. Bennett, N., & Liden, R. (1996). Social exchange in organizations: perceived organizational support, leader-member exchange and employee reciprocity. *Journal of Applied Psychology, 81*, 219-227.

Sherman, M. P. (2004). Antecedents of affective commitment among subgroups of part-time workers: all part-timers are not created equal. Dissertations and Abstracts International: Section B: *The Sciences and Engineering, 65*, (4-B), 2133.

Shore, L. M., & Wayne, S. J. (1993). Commitment and employee behavior: comparison of affective commitment and continuance commitment with perceived organizational support. *Journal of Applied Psychology, 78*, 774-780.

Shore, L. M., & Martin, H. J. (1989). Job satisfaction and organizational commitment in relation to work performance and turnover intentions. *Human Relations, 42*, 625-638.

Stone-Romero, E. F., Stone, D. L., & Salas, E. (2003). The influence of culture on role conceptions and role behavior in organizations. *Applied Psychology: An International Review, 52*, 328-362.

US Census Bureau (2003). Hispanic population in the United States: March 2002. P20-505.

Varona, F. (1996). Relationship between communication satisfaction and organizational commitment in three Guatemalan organizations. *The Journal of Business Communication, 33*, 111-140.

doi:10.1300/J369v09n02_04

Utilizing Task Clarification and Self-Monitoring to Increase Food Temperature Checks Among Restaurant Staff

Rhiannon Fante
Leslie Shier
John Austin

SUMMARY. A within-group reversal design was employed to evaluate the effects of task clarification and self-monitoring on low frequency food preparation behaviors of 16 kitchen employees who worked at one location of a nation-wide restaurant chain. An informal functional assessment was conducted prior to intervention to identify variables responsible for the low frequency of food temperature checks. Based on the assessment results, task clarification and self-monitoring treatments were implemented. The results showed that the mean percentage of appropriate food temperature checks in the first baseline phase was 14%, 17% in the task

---

Rhiannon Fante is a doctoral student in the Department of Psychology, at Western Michigan University, Kalamazoo, MI 49008 (E-mail: rhiannon_fante@hotmail.com).

Leslie Shier, MA, BCBA, is Program Director at The Mariposa School for Children with Autism, 203 Gregson Dr., Cary, NC 27511 (E-mail: leslieshier@yahoo.com).

John Austin, PhD, is Associate Professor, Department of Psychology, Western Michigan University, Kalamazoo, MI 49008 (E-mail: john.Austin@wmich.edu).

[Haworth co-indexing entry note]: "Utilizing Task Clarification and Self-Monitoring to Increase Food Temperature Checks Among Restaurant Staff." Fante, Rhiannon, Leslie Shier, and John Austin. Co-published simultaneously in *Journal of Foodservice Business Research* (The Haworth Hospitality & Tourism Press, an imprint of The Haworth Press, Inc.) Vol. 9, No. 2/3, 2006, pp. 67-88; and: *Human Resources in the Foodservice Industry: Organizational Behavior Management Approaches* (ed: Dennis Reynolds, and Karthik Namasivayam) The Haworth Hospitality & Tourism Press, an imprint of The Haworth Press, 2006, pp. 67-88. Single or multiple copies of this article are available for a fee from The Haworth Document Delivery Service [1-800-HAWORTH, 9:00 a.m. - 5:00 p.m. (EST). E-mail address: docdelivery@haworthpress.com].

Available online at http://jfbr.haworthpress.com

doi:10.1300/J369v09n02_05

clarification phase, 51% in the self-monitoring phase, and 31% in the second baseline phase. The results suggest that task clarification alone, in this situation, was not sufficient to effectively increase performance and self-monitoring is a viable approach to increase low frequency food preparation behaviors such as conducting food temperature checks.
doi: 10.1300/J369v09n02_05 *[Article copies available for a fee from The Haworth Document Delivery Service: 1-800-HAWORTH. E-mail address: <docdelivery@haworthpress.com> Website: <http://www.HaworthPress.com>*

KEYWORDS. Task clarification, self-monitoring, low frequency food preparation behaviors

INTRODUCTION

Customer satisfaction is essential to survival in the restaurant business (Kirwin, 1991). Restaurants and their employees primarily deal with perishable products that are prepared, presented, and consumed by customers (Susskind, Kacmar, & Borchgrevink, 2003). Customers are now, more than ever, demanding efficient and high quality service. A recent study examined the factors that influence a customer's decision to purchase a meal, and found the following factors to be the most critical: expectation level, food temperature, color, and aroma (Hameed, 2002).

According to the National Restaurant Association's 2005 Restaurant Industry Forecast, restaurant industry sales are expected to reach a record of $476 billion in 900,000 restaurant locations in the United States in 2005. The projected annual sales indicate a total economic impact of over $1.2 trillion, and on a typical day the restaurant industry will have average sales of $1.3 billion. These statistics highlight the restaurant industry's critical role in the nation's economy as a job creator. The restaurant industry on average has created about 270,000 new jobs per year during the last 10 years and is expected to add 1.8 million new jobs during the next 10 years (Kim & Dayspring, 2004).

The typical American household spent an average of $2,276 dining out in the year 2002 with the per-capita expenditures on dining averaging $910. The typical quick service and table service restaurant diner is between the age of 30 and 60 years old, is educated, and is more likely to live in larger urban areas. Household spending on dining out is heavily influenced by a variety of demographic characteristics such as: household income, the age of the household head, household size, household composition, number of wage earners, and occupation ("Restaurant Spending," 2002). Given their financial impact and contributions to society, it is clearly worthwhile for restaurants to develop systematic methods to improve the behaviors that lead to exceeding customer expectations, and delivering appropriately cooked food that looks and smells pleasing.

In order for restaurants to thrive, they must produce food that is pleasing to customers. In addition to the business and market factors involved in successfully creating, growing, and managing a restaurant capable of producing pleasing food, there are safety concerns. More specifically, food safety is a major concern in many of the nation's restaurants. It is estimated that foodborne diseases cause approximately 76 million illnesses, 325,000 hospitalizations, and 5,000 deaths in the United States each year (Mead et al., 1999). Known pathogens account for an estimated 14 million illnesses, 60,000 hospitalizations, and 3,309 deaths. Three bacteria pathogens, salmonella, listeria, and toxoplasma are responsible for 1,500 deaths each year, while the other bacteria pathogens combined are responsible for 1,297 deaths each year (Buzby, Roberts, Jordan Lin, & MacDonald, 1996). Parasitic pathogens are responsible for 383 deaths each year and viral pathogens are responsible for 129 deaths each year (Buzby, Roberts, Jordan Lin, & MacDonald, 1996). Unknown agents account for the remaining 62 million illnesses, 265,000 hospitalizations, and 3,200 deaths (Mead et al., 1999). The costs of human illness are estimated to be $9.3-$12.9 billion annually, of these costs, $2.9-$6.7 billion are attributed to foodborne bacteria (Buzby, Roberts, Jordan Lin, & MacDonald, 1996). Given the high frequency and severity (as measured by the number of hospitalizations, deaths, and costs) of events related to foodborne illness, it seem clear that interventions to address these events are warranted.

One approach that can be used to increase food safety practices is an approach based on the principles of applied behavior analysis. Applied behavior analysis is the process of applying the principles of behaviorism to specific behaviors and evaluating whether any changes in the behavior are attributable to that process (Baer, Wolf, & Risley, 1968). Organizational behavior management (OBM) is an area in applied behavior analysis that improves performance by applying the concepts and principles of behaviorism to organizational problems. One OBM approach used to improve the safety behavior of employees is called behavioral safety. The behavioral safety process utilizes the principles of applied behavior analysis and performance management to increase occupational and personal safety (Geller, 2001; Krause 1997, McSween,

1995; Sulzer-Azaroff & Austin, 2000). Behavioral safety techniques are used to motivate employees to improve their performance. These techniques typically involve the following: (1) a careful analysis of the work environment to determine behaviors and situations related to risk; (2) development of a measurement strategy that continues throughout the entire process of addressing the targeted behaviors; (3) analysis of the behaviors and situations to determine their causes; and (4) development and implementation of a treatment that addresses these causes (Sulzer-Azaroff & Austin, 2000).

For example, a study by Geller, Eason, Phillips, and Pierson (1980) was conducted to determine if training and feedback would improve the sanitation practices of cafeteria employees. The intervention targets were responses that increase the probability of microorganisms collecting on hands, and hand washing following the designated microorganism-collecting responses. Microorganism-collecting responses included critical antecedent conditions that should have been followed by hand washing, these included: entering the work area, handling and unpacking food before it had been cleaned, touching the face, touching the hair, touching clothing, touching cleaning equipment, and other. The other category included a variety of low-probability behaviors that should be followed by hand washing (e.g., eating, drinking, touching the floor).

Sanitation training consisted of defining the target behaviors (i.e., task clarification) and sensitizing participants to the importance of reducing the transfer of microorganisms from hands to food. Sanitation training increased the frequency of hand washing on the first workday following training only. The feedback phase consisted of the kitchen supervisor and one of the authors showing individual participants their frequencies of microorganism-collecting and hand washing behaviors during the prior work period. On each day that feedback was administered, the frequency of hand washing more than doubled the highest pretreatment level. Microorganism-collecting responses did not improve with either the sanitation training or feedback.

Although little behavioral research has been conducted on food preparation practices, there is much research on occupational safety and more general customer service practices, and these well-established techniques appear to have some applicability to important food service-related performances. For example, several studies have demonstrated that employee safety behaviors in a wide variety of industries can be increased through the application of behavior management techniques (e.g., Austin, Kessler, Riccobono, & Bailey, 1996; Fox, Hopkins, & Anger, 1987; Sulzer-Azaroff, Loafman, Merante, & Hlavacek, 1990).

Research has also demonstrated the positive effects of behavior management techniques on customer service practices. A study conducted by Brown, Malott, Dillon, and Keeps (1980) was conducted to determine if training and feedback would improve specific customer service behaviors of 3 department store salespeople. The intervention targets were approaching customers, greeting customers, being courteous, and appropriately closing the sale. The 3 participants were exposed to a training program and feedback. The training program had a slight positive impact on service, but performance feedback produced a substantial improvement in the frequency of the targeted behaviors.

Crowell, Anderson, Abel, and Sergio (1988) also conducted a study to determine if behavior management techniques would improve the customer service of bank tellers. The effects of task clarification, feedback, and praise on specific customer service behaviors, which were selected on the basis of extensive preliminary observations of bank teller-customer interactions, were evaluated. The results showed that task clarification produced an overall percentage point increase in desired behaviors of 12% over baseline. Feedback and praise resulted in additional overall increases of 6% and 7%, respectively. A suspension of all procedures resulted in a 9% decrease in overall performance, whereas reinstatement of feedback and praise again resulted in overall increases of 7% and 12% respectively.

Although behavioral techniques such as feedback have been found to be effective and are usually acceptable to employees and employers, these interventions may be impractical for some organizations and occupations. One possible solution is to use self-monitoring to improve employee performance. Self-monitoring approaches are derived from the broad area of self-control strategies. Self-monitoring has been defined as an individual's assessment of whether or not a target behavior has occurred and it is usually followed by self-recording the event (Nelson & Hayes, 1981). Many researchers have studied the effectiveness of self-monitoring as a behavior change strategy, but few researchers have used self-monitoring strategies in organizations.

For example, Calpin, Edelstein, and Redmon (1988) assessed the effectiveness of self-monitoring and self-monitoring plus assigned goals on the proportion of work hours spent in direct client contact by clinicians in a rural mental health center. The results of this study showed that most increases in performance occurred during self-monitoring, but the combined intervention produced additional small increments of improvement in performance.

Olson and Austin (2001) examined the effects of a self-monitoring package on the safety practices of bus operators, which included complete stopping and loading and unloading passengers. The treatment package included self-monitoring, prompts for self-monitoring by the dispatchers, publicly posted group feedback, and occasional ride-alongs from supervisors during which supervisors conducted direct observations of the driver's behavior while on the bus. The intervention increased safe operating behaviors an average of 12.3% (range: 14-41%) when compared to baseline rates. Further improvements in performance occurred on those occasions when supervisors conducted observations.

Hickman and Geller (2003) also examined the effects of a self-monitoring package that included self-management for safety training, self-monitoring with an incentive system, and individual feedback for safe work practices in mining operations. There were 2 conditions in which 8 participants recorded their intentions to engage in specific percentages of safety-related work behaviors before starting their shift for the day (pre-shift monitoring condition) and 7 participants recorded their estimates of their percentages of safety-related work behaviors after their shift for the day (post-shift monitoring condition). The results showed the mean percentage safe score improved by 34.8% (from 34.8% to 46.9%) during intervention for the participants who were in the pre-shift recording condition and the mean percentage safe score improved by 40.1% (from 38.2% to 53.5%) for participants who were in the post-shift recording condition. The results of the studies discussed above suggest that self-monitoring can result in behavior change, but self-monitoring packages may be more effective than self-monitoring alone.

Although the behavioral interventions implemented in the studies discussed above were successful, behavioral researchers may have maximized intervention effectiveness and saved time and resources by using a systematic assessment method. When addressing organizational problems it is common for practioners to conduct a diagnosis (i.e., functional assessment) of the variables that are responsible for maintaining or preventing the behavior. In other words, a functional assessment can help identify the stimulus-response-consequence relationships that are responsible for the target behavior. Research has shown that interventions based on functional assessment methods are more successful than those that are not (Iwata, Pace, Cowdery, & Miltenberger, 1994; Repp, Felce, & Barton, 1988). Although there are three general approaches to conducting functional assessments (informant assessment, descriptive assessment, and experimental analysis) informant assessments are most commonly used by OBM practioners, however OBM practitioners often fail to report doing so in a formal manner (Austin, Carr, & Agnew, 1999). Informant assessment methods are ways to collect information on the variables responsible for maintaining behavior. Most often these methods are informal, such as behavioral interviews, surveys, and rating scales.

One of the first OBM assessment tools published was the Behavior Engineering Model (BEM) created by Gilbert (1978). This model guides the practitioner to ask questions in 6 broad categories: data, instruments, incentives, knowledge, capacity, and motives. Gilbert suggested that this model can aid in the identification of an intervention that will create the most substantial improvement for the organization with the least amount of cost. The BEM provides a global view of performance but lacks the precision of other types of assessment methods.

More recently, Austin (2000) conducted a study to summarize the questions consultants asked when solving organizational problems. He used these summaries to develop an assessment tool called the Performance Diagnostic Checklist (PDC). The PDC guides practitioner questioning and analysis in four main areas: Antecedents and Information, Knowledge and Skills, Equipment and Processes, and Consequences. For example, in the antecedents and information section two sample questions are "Is there a written job description telling exactly what is expected of the employee?" and "Is the supervisor present during task completion?" (p. 340). Two questions from the consequences section are "Do employees see the effects of performance? (How? Natural/arranged) and "Are there other behaviors competing with the desired performance?" The PDC can be used as an interview tool or it can be used to guide the consultant's or manager's assessment by highlighting four areas that should be analyzed.

The PDC has been the most reportedly used assessment tool in OBM research studies in recent years. Rohn, Austin, and Lutrey (2002) conducted an organizational functional assessment, using the Performance Diagnostic Checklist (PDC) (Austin, 2000) as a guide, in order to identify and change variables responsible for maintaining cash shortages in a retail store. An intervention package consisting of feedback and accountability was then implemented based on the results of the assessment. Another study by Pampino, Heering, Wilder, Barton, and Burson (2003) was conducted to examine the utility of the Performance Diagnostic Checklist (PDC) as an assessment tool to design an intervention for increasing maintenance tasks in an independently owned coffee shop. An intervention consisting of task clarification and a lottery were implemented based on the results of the PDC. This package intervention produced an increase in the percentage of closing tasks completed and helped maintain employee performance throughout the intervention period.

Although it is clear that many organizational practitioners conduct an assessment before intervening, maintaining variables are rarely assessed or reported in the literature (Austin, Carr, & Agnew, 1999). The current study employed an intervention package consisting of task clarification and self-monitoring with supervisor prompts to increase low frequency food temperature checks. The components of the intervention package were selected based upon the results of a functional assessment conducted following entry into the organization.

METHOD

Setting and Participants

The present study was conducted at one location of a nation-wide restaurant chain. Normal business hours were Sunday through Thursday 11:00am to 10:00pm and Friday and Saturday 11:00am to 11:00pm. The restaurant's layout consisted of a foyer, several dining rooms, a lounge area, and a kitchen. The kitchen was divided into 5 areas: (a) the dish area, (b) the food preparation area (i.e., temperatures of food are checked, food is placed on trays, etc.), (c) the main cook line (i.e., food is cooked and temperatures of food are checked), (d) the dessert preparation area, and (e) the appetizer preparation area (i.e., soup, salad, etc.).

Participants were sixteen kitchen employees (14 males and 2 females), whose age averaged 21 (range: 18-30). All employees were considered full-time and worked at least 25 to 30 hours every week. Three managers (all males) with experience ranging from 5 to 9 years and an age range of 23-37, helped expedite the project.

Functional Assessment

Prior to intervention, an informal functional assessment was conducted in order to identify some of the causes of low frequency food temperature checks. The Performance Diagnostic Checklist (Austin, 2000) was used as an assessment tool to identify and target specific areas in need of improvement. (See Austin (2000) for the full list of questions and a description of the analysis model used.) The assessment involved answering a series of questions in each of the following four areas: (1) antecedents and information, (2) equipment and processes, (3) knowledge and skills, and (4) consequences. As part of the assessment several interviews were conducted by the primary investigator with each of the managers. The interviews consisted of the primary investigator asking the managers a series of "yes"or "no" questions from the PDC to help identify the areas that were in need of improvement. These areas are reviewed below.

Antecedents and information. The kitchen employees were aware that one of their job responsibilities included taking frequent temperatures of the food after it was prepared. However, employees and managers tended to define "frequent" in different ways. The managers' definition of "frequent" was that food temperatures should be taken several times each hour throughout the entire day. When asked by the primary researcher many of the employees, however, stated that "frequent" simply indicated that they needed to take numerous food temperatures regardless of the time at which the checks occurred.

Recommendations. Because the employees and supervisors reported different performance expectations, a job aid was placed in the employees' immediate environment to clarify expectations. The job aid consisted of a grid of half-hour time blocks, each of which contained a number determined by the managers. This number represented the number of temperature checks that needed to be conducted during that half-hour (see Appendix A for a copy of the job aid).

Equipment and processes. The food temperature was checked (after the food was prepared by the cooks, but before being served to the customer) with a machine called a Quick Check Temperature Gauge. Three temperature gauges were available for employees to use. At the time of the assessment, two primary problems existed with these machines: (a) the temperature gauges were frequently broken or not working properly, and (b) the gauges were not easily accessible to the employees. Many times the employees had trouble finding them, or could not find them at all. Because there was no designated place for the temperature gauges they were often left on the counter where they interfered with other additional duties, including placing food dishes on trays to be served, preparing appetizers, and preparing desserts, making the gauges more susceptible to damage.

Recommendations. In order to eliminate the problem of the temperature gauges not being optimally arranged in the employees' environment, the manager, along with the employees who used the temperature gauges most frequently, agreed on a storage location for them. A check sheet was created for the managers, so they could track which days the temperature gauges were or were not in proper working order. In other words, the managers calibrated each temperature gauge to ensure they were giving accurate temperature readings. There was also a section on the check sheet for the managers to verify that the temperature gauges were in their proper locations.

Knowledge and skills. Employees, whose responsibilities included checking food temperatures, received on-the-job training on how to use the temperature gauge at the start of their employment. Each employee appeared to have the knowledge and skills required to correctly check food temperatures.

Recommendations. There were no recommendations involving knowledge and skills because no deficiencies were identified in these areas.

Consequences. Consequences contingent upon task completion (i.e., required number of food temperature checks conducted) or upon non-completion (i.e., no and/or very few food temperature checks conducted) were seldom delivered. Feedback, if given, usually occurred at least a day or more after task completion, since performance monitoring was conducted by a computer and computer printouts were seldom reviewed by the managers. All temperatures checked for each day were downloaded into a computer program, so neither the employees nor managers were aware of the employees' performance level on an ongoing basis.

Recommendations. Since performance feedback from computer monitoring was delayed, a self-monitoring form was created so that all employees could track their own performance in real time. In addition, managers were instructed to prompt employees' use of the self-monitoring forms on a daily basis.

Assessment Results

The functional assessment identified deficiencies in the "antecedents and information" category, the "equipment and processes" category and the "consequences" category. Based on the assessment an intervention consisting of a task clarification and self-monitoring, in which managers prompted self-monitoring, was designed and implemented.

Procedure and Design

Traditional methods of evaluating the effects of an intervention rely upon the random assignment of participants into experimental groups. These methods however, are not feasible in most industrial settings, since randomization cannot be arranged in most work environments.

An alternative to this approach is the use of a within-group experimental design. Within-group designs have been used by the majority of behavioral researchers assessing the effectiveness of interventions in work settings (Lingard & Rowlinson, 1997). This study utilized a within-group reversal design, which involves the drawing of comparisions within the same group of participants by taking repeated and frequent measures over time, some when participants are exposed to the independent variable and others when they are not exposed to the independent variable. In such a design, an effect is demonstrated by measuring behavior during baseline (A1) and then producing behavior change only when the treatment (B1) is applied. Experimental control is demonstrated only when removing the treatment causes the behavior to return to baseline (B2). The experimental effect and control is said to be *replicated* (i.e., bolstered) when the treatment is readministered a final time (B2). Although we did not attempt this in this study, further manipulations (e.g., ABABABA) of the independent variable would be expected to produce similar effects on the dependent variable and would be even stronger evidence of experimental control (Kazdin, 1982).

This study evaluated the effects of task clarification and self-monitoring on food temperature checks. Data were collected every day via computer linked to the Quick Check temperature gauges and were picked up once every week by the primary investigator. Data were entered into a spreadsheet programmed to calculate the percentage of half-hour time blocks in which the minimum number of temperature checks were conducted throughout each day of the week. Employees were not informed of the research study or the role of the primary investigator until after the study was completed.

Baseline. Because the managers had computer printouts on the number of temperature checks for the previous two and half months prior to the study baseline data were collected for 89 days prior to the introduction of intervention.

Task clarification. On day 90, two job aids that indicated how many temperature checks the employees need to complete every half hour, were placed in the employees' immediate environment. Before each shift during this phase the managers explained the purpose of the job aid to make sure that each employee understood that he or she was expected to conduct the specified number of temperature checks for every half hour of the shift. Two job aids were posted, because employees checking food temperatures worked in two different areas of the kitchen. This phase lasted for 8 days.

Self-monitoring. On day 98, the job aids were removed and self-monitoring forms were introduced. The self-monitoring forms were very similar in appearance to the job aid. The only difference was that on the self-monitoring forms employees were expected to circle a number corresponding to the number of checks completed during that time block. By circling a number after each check the employees were able to see how many temperature checks they had completed, along with how many temperature checks were needed to meet the required number for a particular time block (see Appendix B for a sample of the self-monitoring sheet).

The managers explained the self-monitoring forms to all of the employees. In addition the managers prompted the employees' use of the form to ensure that the forms were being regularly completed. Prior to the introduction of the self-monitoring forms, there was no convenient way to make sure that the employees had the opportunity (i.e., that enough food was ordered during that time) to meet the criterion specified for each half-hour time block. However, each time block on the self-monitoring form allowed the employees to circle N/A (non-applicable) if they could not meet the specified criterion because not enough food was prepared during the time interval.

Return to baseline. On day 105 the self-monitoring forms were removed and the intervention was withdrawn for nine days. As a result, the employees' environment appeared as it had during the initial baseline period.

Interobserver Agreement

All temperatures checked for each day, using the temperature gauges were downloaded into a computer and then were printed out at the end of the week. The temperatures recorded on the computer printout were then entered, by the primary investigator, into a spreadsheet programmed to calculate the percentage of half-hour time blocks meeting the minimum number of temperature checks for each day. Because human error can occur during data entry, interobserver agreement was conducted for 50% of the data entry. Interobserver agreement was calculated by comparing the primary investigator's data entry by a second investigator's data entry and by dividing the number of agreements by the number of agreements plus disagreements. An agreement was counted when both data sheets had a half-hour time block that met the specified criterion. A disagreement was counted when one data sheet had a time block that met the specified criterion while the same block on the second data sheet did not meet the specified criterion. Interobserver agreement averaged 98% throughout the course of the study.

Independent Variable Integrity

Steps were taken to ensure that both components of the intervention were implemented as planned. The primary investigator was in the manager's presence when both job aids were posted and when the manager explained the purpose of the job aid to the employees. In order to make sure that the job aids remained posted, the primary investigator visited the business a few days after they were initially posted to verify that they were in fact in their designated locations. To ensure integrity of the self-monitoring procedure the primary investigator cross-checked the self-monitoring forms with the computer printouts. The managers were also asked to prompt the participants to complete the self-monitoring forms when they saw the participants failing to record a temperature check on the self-monitoring form.

Independent variable integrity was also calculated to determine the percentage of occasions on which employees used the self-monitoring forms. The primary investigator calculated the number of times the employees used the self-monitoring forms; this number was attained by dividing the total number of times the employees circled the numbers on the self-monitoring forms divided by the total number of opportunities to circle a number on the form. The employees used the self-monitoring form 21% of the time during the self-monitoring phase.

Employees were required to circle a number on the self-monitoring form corresponding to the number of checks they completed during each time block. Since the self-monitoring forms depicted time blocks for both the lunch and dinner shift, one of the employees working the dinner shift turned in the self-monitoring form at the end of the night to the manager. The self-monitoring forms were collected weekly by the primary investigator along with the computer printouts. The primary investigator then used the computer printouts and not the self-monitoring forms to graph the primary dependent variable and report the results of this study.

RESULTS

Baseline. During the first baseline phase, the average percentage of time blocks in which the employees met the temperature checking criteria was 14.3% (*SD* = 13.8%, range = 0%-61.5%).

FIGURE 1. Percentage of Half-Hour Time Blocks That Met the Temperature Check Criterion. The Graph Depicts Data for All Employees Combined

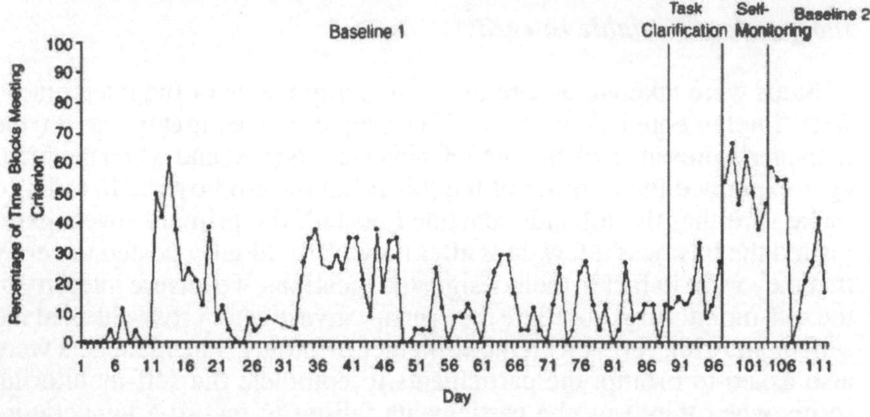

Task clarification phase. After the implementation of the two job aids the average percentage of time blocks in which the employees met the temperature checking criteria was 17.1% (SD = 8.5%, range: 8.3%-33.3%), an increase of 3 percentage points over baseline.

Self-monitoring phase. Implementing the self-monitoring form along with supervisor prompting produced an immediate increase in the time blocks in which the employees met the temperature checking criteria. The average performance during the self-monitoring phase was 51.7% (*SD* = 10.1%, range: 37.5%-66.6%), an increase of 34 percentage points over the previous phase.

Return to baseline. During the second baseline phase, the average percentage of time blocks in which the employees met the temperature checking criteria was 31.2% (*SD* = 21.5%, range: 0%-58.3%). This represented an average decrease of 20.5 percentage points over the previous phase.

Although the graph shows and increase in the average percentage of time blocks in which the employees met the temperature checking criteria during the self-monitoring phase, effect sizes for each phase were calculated in order to facilitate a closer examination of the data. The effect size is a measure of the magnitude of an effect, and the effect size calculations in the current study were performed using the equation presented by McConville, Hantula, and Axelrod (1998). As a reference for interpreting effect sizes, Cohen (1988) stated that effect sizes of .2-.49 should be considered small; .5-.79 should be considered medium, and effect sizes of .8 or greater should be considered large. Table 1 provides a summary of the effect sizes observed for each phase, in comparison to baseline performance.

DISCUSSION

The results of this study suggest that task clarification alone was not effective in improving performance but that adding self-monitoring was sufficient to increase low frequency food temperature checks. During the task clarification phase, on average, there was only a 3 percentage point increase in performance above the baseline mean. However, the addition of self-monitoring along with supervisor prompts to engage in self-monitoring resulted in a more substantial mean increase of 37 percentage points above the baseline mean in the target performance.

One possible explanation for these results is that the self-monitoring forms reminded employees to conduct checks and that the act of self-monitoring allowed each employee to keep track of his or her own performance in real time. The employees may then have increased the frequency of checking food temperatures as a result of this increased prompting and awareness of their performance level, which was in the past not regularly reviewed with them. As a result, the employees may then have made an effort to meet the criterion that was set for their shift. A second possible explanation is that when the supervisors prompted employee use of the self-monitoring forms, this also prompted employees to check food temperatures. While it is clear that the intervention improved the participants' performance it is not possible to distinguish which part(s) of the intervention (reactivity to managerial prompting and/or the effects of the self-monitoring) was responsible for the improvement.

The purpose of the functional assessment conducted prior to the implementation of the intervention was to identify variables responsible for the

TABLE 1. The Overall Effect Size for Each Condition (In Relation to Baseline Performance)

Experimental Condition		
Task Clarification	**Self-Monitoring**	**Return to Baseline**
0.2	2.75	1.15

low frequency of behavior. The limited effects of the task clarification phase and the success of the self-monitoring phase suggests that conducting such an assessment was effective in identifying at least one of the variables responsible for maintaining behavior. Further studies could be conducted to evaluate the effectiveness of an intervention based on the results of a functional assessment compared to an arbitrary intervention.

Although the self-monitoring phase was effective at increasing the frequency of checking food temperatures, there are some methodological limitations that apply to the current study and more generally to self-monitoring treatments. The difficulties in ascertaining reliability of self-report for events which have no verification, and the reactivity of self-monitoring limit the value of such interventions (Kanfer, 1970). In the current study, the employees marked *non-applicable* on the self-monitoring form if the opportunity to meet the criterion was not available for a particular time block. A problem with this method was that there was no means to verify whether the participant did have the opportunity and did not engage in the desired behavior or whether he or she did not have the opportunity to engage in the behavior. For this reason, if the computer printout indicated that the required number of temperature checks was not met and non-applicable was circled on the self-monitoring form, the employees were scored on the dependent variable as not having met the criterion for that half-hour. This likely led to a more conservative estimate of the treatment effects.

A study conducted by Critchfield (1999) found that the frequency of self-monitoring can have an effect on its impact. Since employees were instructed to engage in self-monitoring after every occurrence of the target behavior this intervention was very high in response effort and may have been more intrusive than necessary. A less intrusive self-monitoring procedure, such as a self-monitoring form requiring monitoring estimates at the beginning or the end of the day, may have been less effortful, and may have produced the same results as those that were produced with this more intrusive procedure.

Even though the self-monitoring phase resulted in an increase in performance it is not possible to determine whether the self-monitoring, the supervisor prompts to monitor, or some combination of the two were responsible for the increase in performance. Additional studies could be conducted on the self-monitoring component alone since there are methodological limitations (e.g., the difficulty in ascertaining reliability of self-report and reactivity to self-monitoring) when using self-monitoring. Future studies could help to develop more original and creative ways to ascertain reliability of self-reported events, such as corroboration with observations by a trained researcher or by video observations. In addition, more research is needed on the extent to which accuracy and frequency of monitoring influences the effectiveness of self-monitoring.

Despite the limitations of the current study mentioned above, the primary data collection indicated the procedure was successful, and anecdotal reports gathered suggest that the managers and employees believed the intervention was effective, easy to implement, and helpful in freeing up time for the employees to perform other responsibilities. The current study also suggests that using an intervention based on the results of a functional assessment is an effective approach that can be used by managers and practitioners to solve performance problems.

This study successfully used an OBM based strategy to change the behavior of a restaurant staff. Every aspect of business and industry involves behavior and improving any important outcome involves changing behavior. OBM provides managers and leaders with scientifically-validated tools to create and sustain behavior change in any business or industry. In order to use these tools effectively managers must understand that behavior is a function of the environment in which it occurs. Ineffective managers create a work environment in which employees perform only enough to meet the minimal requirements to keep their job, whereas effective managers create an environment in which employees want to perform above the minimum requirements (Daniels & Daniels, 2004).

Managers can use the tools of OBM to create a work environment in which employees want to perform well. With OBM managers can learn to effectively: operationally (i.e., clearly) define behavior, monitor work performance, change consequences to support behavior, and remove performance barriers (Daniels & Daniels, 2004). This will help managers think about what outcomes they want employees to produce and what employees need to do to produce those outcomes.

Research in OBM has shown that effective managers stay better informed about their employees' performance through work monitoring (Komaki, 1998). Work monitoring includes reviewing results and work processes, discussing work performance with employees, and listening to employees concerns and suggestions before providing performance feedback based on the data collected during monitoring. OBM also teaches managers how to improve employee performance and help understand why employees do what they do by analyzing the consequences that follow behavior. Typical managerial strategies such as reminding, nagging, and various methods of persuasion at first seem to change employee behavior, but in fact produce little lasting behavior change. In addition to ineffective consequences, managers often must remove barriers that prevent ideal performance. Effective mangers need to consider ways to change the job to make it easier and more convenient for employees to perform better.

The success of an organization depends on the effectiveness of its processes, the effectiveness of its processes depends on the behavior of its employees, and the behavior of its employees depends on the skill of its managers. When successfully applied, OBM techniques help managers to identify and monitor employee behaviors, encourage productive behaviors, and eliminate barriers to those behaviors.

REFERENCES

Austin, J. (2000). Performance analysis and performance diagnostics. In J. Austin & J. Carr (Eds.), *Handbook of applied behavior analysis* (pp. 304-327). Reno, Nevada: Context Press.

Austin, J., Carr, J. E., & Agnew, J. A. (1999). The need for measures of maintaining variables in OBM. *Journal of Organizational Behavior Management, 19*(2), 73-90.

Austin, J., Kessler, M. L., Riccobono, J .E., & Bailey, J. S. (1996). Using feedback and reinforcement to improve the performance and safety of a roofing crew. *Journal of Organizational Behavior Management, 16*(2), 49-75.

Baer, D. M., Wolf, M. M., & Risley, T. R. (1968). Some current dimensions of applied behavior analysis. *Journal of Applied Behavior Analysis, 1*, 91-97.

Brown, M. G., Malott, R. W., Dillon, M. J., & Keeps, E. J. (1980). Improving customer service in a large department store through the use of training and feedback. *Journal of Organizational Behavior Management, 2*(4), 251-265.

Buzby, J. C., Roberts, T., Jordan Lin, C. T., & MacDonald, J. M. (1996). Bacterial foodborne diseases: Medical costs and productivity losses. *Agricultural Economics Report, 741* 100. Retrieved February 5, 2005, from http://www.ers.usda.gov/publications/aer741/

Calpin, J. P., Edelstein, B., & Redmon, W. K. (1988). Performance feedback and goal setting to improve mental health center staff productivity. *Journal of Organizational Behavior Management, 9*(2), 35-58.

Critchfield, T. S. (1999). An unexpected effect of recording frequency in reactive self-monitoring. *Journal of Applied Behavior Analysis, 32*, 389-391.

Crowell, C. R., Anderson, C. D., Abel, D. M., & Sergio, J. P. (1988). Task clarification, performance feedback and social praise: Procedures for improving the customer service of bank tellers. *Journal of Applied Behavior Analysis, 21*, 65-71.

Daniels, A. C., & Daniels, J. E. (2004). *Performance management: Changing behavior that drives organizational effectiveness*. Atlanta, GA: Aubrey Daniels International, Inc.

Fox, D. K., Hopkins, B. L., & Anger, W. K. (1987). The long-term effects of a token economy on safety performance in open-pit mining. *Journal of Applied Behavior Analysis, 20*, 215-224.

Geller, E. S. (2001). *Working safe: How to help people actively care for health and safety*. Boca Raton, FL: Lewis Publishers.

Geller, E. S., Eason, S. L., Phillips, J. A., & Pierson, M. D. (1980). Interventions to improve sanitation during food preparation. *Journal of Organizational Behavior Management, 20*(3), 229-240.

Gilbert, T. F. (1978). *Human competence*. Amherst, MA: HRD Press, Inc.

Hameed, A. (2002). Factors influencing food preference decisions of restaurant clients. *Administrative Sciences, 29*, 424-444.

Hickman, J. S., & Geller, E. S. (2003). A safety self-management intervention for mining operations. *Journal of Safety Research, 34*, 299-308.

Iwata, B. A., Pace, G. M., Kalsher, M. J., Cowdery, G. E., & Cataldo, M. F. (1990). Experimental analysis and extinction of self-injurious escape behavior. *Journal of Applied Behavior Analysis, 23*, 11-27.

Kanfer, F. H. (1970). Self-monitoring: Methodological limitations and clinical applications. *Journal of Consulting and Clinical Psychology, 35*, 148-152.

Kazdin, A. E. (1982). Single-case experimental designs in clinical research and practice. *New Directions for Methodology of Social & Behavioral Science, 13*, 33-47.

Kim, K., & Dayspring, B. (2004, December). National restaurant association announces record sales projected in year ahead for nation's largest private-sector employer. *The Restaurant Industry Forecast, 36*. Retrieved February 5, 2005, from http://www.restaurant.org/pressroom/pressrelease.cfm?ID = 979.

Kirwin, P. (1991). The satisfaction of service. *Lodging Hospitality, 47*, 66.

Komaki, J. L. (1998). *Leadership from an operant perspective*. New York, Routledge.

Krause, T. R. (1997).*The behavior-based safety process*. New York, Van Nostrand Reinhold.

Lingard, H. & Rowlinson, S. (1997). Behavior-based safety management in Hong Kong's construction industry. *Journal of Safety Research, 28*, 243-256.

McConville, M. L., Hantula, D. A., & Axelrod, S. (1998). Matching training procedures to outcomes: A behavioral and quantitative analysis. *Behavior Modification*, 22(3), 391-414.

McSween, T. M. (1995). *The values-based safety process*. New York, Van Nostrand Reinhold.

Mead, P. S., Slutsker, L., Dietz, V., McCaig, L. F., Bresee, J. S., Shapiro, C., Griffin, P. M., & Tauxe, R. V. (1999). Food-related illnesses and death in the United States. *CDC Emerging Infectious Diseases*, 5. Retrieved February, 5 2005, from http://www.cdc.gov/ncidod/eid/vol5no5/mead.htm.

Nelson, R. O., & Hayes, S. C. (1981). Theoretical explanations for reactivity in self-monitoring. *Behavior Modification, 5*, 3-14.

Olson, R., & Austin, J. (2001). Behavior-based safety and working alone: The effects of a self-monitoring package on the safe performance of bus operators. *Journal of Organizational Behavior Management, 21*(3), 5-43.

Pampino, Jr., R. N., Heering, P. W., Wilder, D. A., Barton, C. G., & Burson, L. M. (2003). The use of the performance diagnostic checklist to guide intervention selection in an independently owned coffee shop. *Journal of Organizational Behavior Management, 23*(2/3), 5-19.

Repp, A. C., Felce, D., Barton, L. E., (1988). Basing the treatment of stereotypic and self-injurious behaviors on hypotheses of their causes. *Journal of Applied Behavior Analysis, 21*, 281-289.

Restaurant Spending: Consumer Expenditure Survey. (2002). Retrieved February 5, 2005, from http://www.restaurant.org/research/consumer/spending.cfm.

Rohn, D., Austin, J., & Lutrey, S. M. (2002). Using feedback and performance accountability to decrease cash register shortages. *Journal of Organizational Behavior Management, 22*(1), 33-46.

Sulzer-Azaroff, B., & Austin, J. (2000). Behavior-based safety and injury reduction: A survey of the evidence. *Professional Safety, 45*(7), 19-24.

Sulzer-Azaroff, B., Loafman, B., Merante, R. J., & Hlavacek, A. C. (1990). Improving occupational safety in a large industrial plant: A systematic replication. *Journal of Organizational Behavior Management, 11*(1), 99-120.

Susskind, M., Kacmar, M., & Borchgrevink, C. P. (2003). Customer service providers' attitudes relating to customer service and customer satisfaction in the customer-server exchange. *Journal of Applied Psychology, 88*, 179-187.

doi:10.1300/J369v09n02_05

Appendix A

Temperature Check Job Aid

Lunch **Number of Temperatures**

10:00-10:30am	10:30-11:00am	11:00-11:30am	11:30-12:00pm	12:00-12:30pm	12:30-1:00pm	1:00-1:30pm	1:30-2:00pm	2:00-2:30pm
1	1	4	10	10	10	10	7	7

2:30-3:00pm	3:00-3:30pm	3:30-4:00pm
5	5	5

Dinner

4:00-4:30pm	4:30-5:00pm	5:00-5:30pm	5:30-6:00pm	6:00-6:30pm	6:30-7:00pm	7:00-7:30pm	7:30-8:00pm	8:00-8:30pm
1	1	4	10	10	10	10	7	7

8:30-9:00pm	9:00-9:30pm	9:30-10:00pm	10:00-10:30pm	10:30-11:00pm
5	5	5	5	5
			Friday and Saturday Only	

Directions

(1) Each block represents one half-hour period.
(2) The number in each block is the number of temperatures that need to be taken for that half-hour.
(3) Once that number is reached for that time block no more temperatures for that half-hour need to be taken.
(4) If there is a half-hour time block where there is no food to be temped than another item needs to be temped (for example soup).

Appendix B

Temperature Check Self-Monitoring Form

Lunch **Number of Temperatures**

10:00-10:30am	10:30-11:00am	11:00-11:30am	11:30-12:00pm	12:00-12:30pm	12:30-1:00pm	1:00-1:30pm	1:30-2:00pm	2:00-2:30pm
1 N/A	1 N/A	1 2 3 4 N/A	1 2 3 4 5 6 7 8 9 10 N/A	1 2 3 4 5 6 7 8 9 10 N/A	1 2 3 4 5 6 7 8 9 10 N/A	1 2 3 4 5 6 7 8 9 10 N/A	1 2 3 4 5 6 7 N/A	1 2 3 4 5 6 7 N/A

2:30-3:00pm	3:00-3:30pm	3:30-4:00pm
1 2 3 4 5 N/A	1 2 3 4 5 N/A	1 2 3 4 5 N/A

Dinner

4:00-4:30pm	4:30-5:00pm	5:00-5:30pm	5:30-6:00pm	6:00-6:30pm	6:30-7:00pm	7:00-7:30pm	7:30-8:00pm	8:00-8:30pm
1 N/A	1 N/A	1 2 3 4 N/A	1 2 3 4 5 6 7 8 9 10 N/A	1 2 3 4 5 6 7 8 9 10 N/A	1 2 3 4 5 6 7 8 9 10 N/A	1 2 3 4 5 6 7 8 9 10 N/A	1 2 3 4 5 6 7 N/A	1 2 3 4 5 6 7 N/A

8:30-9:00pm	9:00-9:30pm	9:30-10:00pm	10:00-10:30pm	10:30-11:00pm
1 2 3 4 5 N/A	1 2 3 4 5 N/A	1 2 3 4 5 N/A	1 2 3 4 5 N/A	1 2 3 4 5 N/A
			Friday and Saturday Only	

Directions

(1) Each block represents one half-hour period.
(2) The numbers in each block are the number of temperatures that need to be taken for that half-hour.
(3) You need to circle each number as you take temperatures (e.g., at 6:30pm you have taken 6 all numbers up to 6 need to be circled)
(4) N/A means that there was not enough food cooked to reach that many temperatures.
(5) However if some temperatures were taken but you do not have enough food to reach all numbers the previous numbers should be circled.
(6) Once that number is reached for that time block no more temperatures for that half-hour need to be taken.

Date: ____________

Why Restaurant Sales Contests Are Self-Defeating

David L. Corsun
Amy L. McManus
Clark Kincaid

SUMMARY. Restaurant sales contests (RSCs) have no substantive body of literature to confirm or disconfirm their effectiveness. In U.S. organizations overall, sales contest (SC) popularity has been documented for decades, increasing in overall expenditure. In U.S. restaurants, the literature notes little more than that SC initiatives are widespread in current use, remaining popular throughout the past decade. However, serious concerns surface when examining the basic tenets that underlie RSC use and the lack of generalizability in SC studies of the business disciplines. In an attempt to open a wide area of fruitful potential research benefiting both theory and practice, we critically analyze the topic of RSCs from a scholarly and practitioner-based perspective, presenting propositions that integrate theoretical and empirical SC works with

---

David L. Corsun, PhD, is Associate Professor, William F. Harrah School of Hotel Administration, University of Nevada, Las Vegas, 4505 Maryland Parkway, Box 456021, Las Vegas, NV 89154-6021 (E-mail: dcorsun@ccmail.nevada.edu).

Amy L. McManus is Doctoral Student, University of Nevada, Las Vegas.

Clark Kincaid, PhD, is Assistant Professor, University of Nevada, Las Vegas.

*Please contact the first author with all inquiries.

[Haworth co-indexing entry note]: "Why Restaurant Sales Contests Are Self-Defeating." Corsun, David L., Amy L. McManus, and Clark Kincaid. Co-published simultaneously in *Journal of Foodservice Business Research* (The Haworth Hospitality & Tourism Press, an imprint of The Haworth Press, Inc.) Vol. 9, No. 2/3, 2006, pp. 89-109; and: *Human Resources in the Foodservice Industry: Organizational Behavior Management Approaches* (ed: Dennis Reynolds, and Karthik Namasivayam) The Haworth Hospitality & Tourism Press, an imprint of The Haworth Press, 2006, pp. 89-109. Single or multiple copies of this article are available for a fee from The Haworth Document Delivery Service [1-800-HAWORTH, 9:00 a.m. - 5:00 p.m. (EST). E-mail address: docdelivery@haworthpress.com].

Available online at http://jfbr.haworthpress.com

doi:10.1300/J369v09n02_06

current understandings of restaurant organizations and their human resource objectives. doi:10.1300/J369v09n02_06

KEYWORDS. Sales contests, motivation, total quality management, front-of-the-house, restaurant

> According to numerous studies in laboratories, workplaces, classrooms, and other settings, rewards typically undermine the very processes they are intended to enhance. The findings suggest that the failure of any given incentive program is due less to a glitch in the program than to the inadequacy of the psychological assumptions that ground all such plans. (Kohn, 1993, p. 54)

> A poorly planned sales contest fails before it begins. In addition, a poorly planned sales contest can cause many more problems than it can solve ... The results, under these circumstances, always seem to be less than satisfactory. (Moncrief, Hart & Robertson, 1988, p. 56)

INTRODUCTION

Sales contests (SCs) are a widely popular incentive used by restaurant organizations to increase check average and total revenue, amongst other goals and in a number of weakly-related guises. Sales contests are programs used to compensate and/or motivate salespersons on a short-term basis to meet specific organizational objectives (Churchill, Ford & Walker, 1993; Futrell, 1981; Murphy & Sohi, 1995), and their use across several industries has increased at a rapid rate over the past two decades (Murphy, Dacin, & Ford, 2004). U.S. expenditures on sales contests totaled $4 billion in the 1980s, again doubling by the mid-1990s (Chrapek, 1989; Nolan & Alonzo, 1997; in Murphy, Dacin & Ford, 2004).

This parallels the expansion of SC literature during this time, which would serve to inform practice. Contests have been repeatedly recommended to managers (e.g., Beltramini & Evans, 1988) as they have been found to benefit contest related goals (Wotruba & Schoel, 1983) while not negatively affecting the sales of unrelated products (Wildt, Parker & Harris, 1987). Overall, SC related research foci include–but are not limited to–performance, motivation, economics, finance and marketing (e.g., Coughlan & Sen, 1989). In the area that has received the most attention–motivation–researchers have applied a variety of theories, such as agency theory (e.g., Bartol, 1999; Basu, Srinivisan & Staelin, 1985; Ghosh & John, 2000; Ross, 1973), transaction cost analysis (John & Weitz, 2001), goal-setting and alignment of objectives (e.g., Farley, 1964; Hart, 1985; Joseph & Kalwani, 1998), and related issues of risk and uncertainty (Harris & Raviv, 1979; Lal & Staelin, 1986; Pratt, 1964; Shavell, 1979).

Despite these research efforts, however, SC practices remain relatively unchanged both in restaurants and in other industries. In restaurants, commonly recommended restaurant sales contests (RSCs) include, but are not limited to: rewards for guest checks covering all categories of menu items, low ticket times, sales per hour, overall revenue, and highest team check averages (e.g., Berta, 2004; Cebrzynski, 2002; Marvin, 1997; Sullivan, 2002). SC goals are typically product-, revenue-, or profit margin-focused (e.g., specific ingredients, such as premium instead of well liquors; and specific item categories, such as appetizers or desserts); rather than customer-focused. This conflict in foci calls into question whether progress in the general business literature is applicable to a restaurant context, as research findings seem to have had little to no impact on how RSCs have been conducted over the years.

We must reexamine the use of restaurant sales contests (RSCs) for the following reasons. First, research on sales contests has sampled a wide variety of occupations and organizational contexts; but none that compare to front-of-the-house restaurant and food servers. Second, other studies on salesforce compensation make various assumptions that do not always hold true in restaurant organizations because of the nature of the experiential product created. Third, and possibly most important, sales contests reward one behavior: selling. Motivating employees toward this goal with a reward moves the focus away from service, which is often the core competency of restaurant organizations directed toward long-term performance.

In this article, we first describe SCs, their typologies, objectives, and effectiveness as described in the literature. Second, we examine the SC literature for evidence of generalizability to the restaurant context. Next, we critique the intent, implementation, and process of RSCs and present propositions based on related literature. Last, we search for a holistic model of RSCs and explore the results of applying Deming's Total Quality Management (TQM) model to a hypothetical RSC.

SALES CONTEST OBJECTIVES

Organizations employ sales contests for a variety of objectives, including "increasing overall sales, increasing market penetration, introducing new products, overcoming seasonal slumps, and easing unfavorable inventory situations" (Kalra & Shi, 2001, p. 171). In an attempt to reach these objectives, a vast amount of research has focused on the optimal designs of contests, allocation of rewards, and rank-order differences in reward (Churchill, Ford & Walker, 1993; Kalra & Shi, 2001; Kotler, 1997); as well as the comparative fit of sales contests versus traditional piece-rate compensation (Lazear & Rosen, 1981, Green & Stokey, 1983; Nalebuff & Stiglitz, 1983; in Kalra & Shi, 2001).

In most cases, the primary objective of sales contests is to generate higher sales volume and net profit (Wildt, Parker & Harris, 1980-1981). With this primary goal, SCs in the restaurant context are redundant, at least for those employees whose income derives largely from gratuities. For this group, which includes nearly all servers in the U.S. and in other countries, income increases in approximately direct proportion with sales volume. The sales contest thus serves to act as a secondary form of direct compensation for increasing sales. The major difference between the rewards derived from increased tips and RSCs is that all servers have access to increased tips, whereas access to the compensation arising from RSCs is available only to the contest winner.

Further, actions servers take outside of increased selling to increase their customers' satisfaction–such as increasing familiarity with patrons (e.g., Crusco & Wetzel, 1984; Garrity & Degelman, 1990; Lynn & Mynier, 1993)–tend to increase their compensation through increased gratuities (Mok & Hansen, 1999). Previous research suggests that perceived quality of service may be more highly correlated with tip percentage than the size of the bill (Boyes, Mounts & Sowell, 2004)–a reward for behaviors that increase the long-term profitability of a restaurant. Thus the objectives of the restaurant and the server are closely aligned and jointly understood, as there may be a higher monetary reward for "serving" than for "selling."

SALES CONTEST TYPOLOGIES AND REWARD ALLOCATION

There are two typologies of sales contest approaches, each of which has three categories and all of which involve some form of competition. The first typology (Moncrief, Hart & Robinson, 1988) includes the following categories: (1) an individual versus other salespeople, (2) a team versus other teams, and (3) an individual versus his or her quota. Organizations and individual employees differ in their preferences for one approach versus another, thus managers have been advised to determine which approach is best suited to their particular employee base and the goals of the incentive program before implementation (Moncrief, Hart & Robinson, 1988).

The second typology (Kalra & Shi, 2001) is divided into two schemes: rank order and winner-takes-all. Rank order compensates employees on their relative performance outcomes, in theory removing common sales uncertainties from the equation. Winner-takes-all, self-explanatory in nature, compensates the top winner only. Winners can be either individual, team, or other-group-based in designs composing this typology.

The literature provides evidence of variability in three broad areas: the cause of SC failure, the distribution of rewards, and which contest characteristics are liked the least by individuals and based on job context. The causes of SC failure are numerous, and will be discussed at length to follow; but in general, contest incentives that compensate participants may impede the intended outcomes by motivating the wrong behavior (Kohn, 1993; Kurland, 1995).

Distribution of rewards is also of interest to the restaurant context, since disproportionate SC wins by heterogeneous employee groups have disadvantages. High performers have been shown to be significantly more sensitive to fluctuations in compensation, and their satisfaction as predicted by total compensation and inequity perceptions has been found to be twice as strong as lower performers (R-squared [high] = 0.42; R-squared [low] = 0.19; Joseph & Kalwani, 1998). If this phenomenon exists among restaurant servers participating in RSCs, a winner-takes-all design may tend to cause dissatisfaction and inequity to a higher degree with the higher performers–punishing high-performing non-winners more than low-performing non-winners.

A third area of inquiry explores the individual differences in salesperson preferences for certain contest characteristics. The generally accepted rule is that approximately 40% of the sales force should receive some type of reward (Churchill, Ford & Walker, 1985) and that there should be different reward levels (Moncrief, Hart & Robinson, 1988). Lack of fairness in rewards was also listed as the number one reason for SC failure in the lattermost study.

EVIDENCE OF SALES CONTEST EFFECTIVENESS

Decades after sales contests have been in ample use, confirmed evidence of their effectiveness is still relatively weak (Moncrief, Hart & Robertson, 1988; Wildt, Parker & Harris, 1980; Wotruba & Schoel, 1983). Sales contests differ from many other forms of compensation in that they are not part of a regular compensation package, and aim at short-term outcomes (Moncrief, Hart & Robinson, 1988). They pay in relative, rather than absolute terms (Kalra & Shi, 2001), and are often separated into categories that vary by researcher. Thus, building sales contest theory upward from the general sales force compensation literature may be inappropriate.

Sales contests have been criticized for their success only in gaining temporary compliance (Kohn, 1993), which disappears once rewards are absent: "When it comes to producing lasting change in attitudes and behavior ... rewards, like punishment, are strikingly ineffective. Once the rewards run out, people revert to their old behaviors" (p. 55). Industry professionals have also observed that "contests and incentives are narcotic in nature–the more you use them, the more you need to use them" (Cichelli, in Marchetti, 2004, p. 19).

The negative possible effects of sales contests, even those well-designed, include inappropriate participant behaviors such as manipulating sales timing, pushing contest-related products to an improper degree, and decreasing cooperation among coworkers (Moncrief, Hart & Robinson, 1988; Wildt, Parker & Harris, 1980-1981; Wotruba & Schoel, 1983). Other concerns raised are employee morale, decreased or redirected effort away from other duties, employee expectations for constant rewards, and decreased performance following contests (Moncrief, Hart & Robertson, 1988). Still more studies discuss adverse impacts on general sales volume, motivation of salespeople, overall side effects for the organization, and customer relationships (e.g., Wildt, Parker & Harris, 1980-1981). An empirically-based foundation of these issues has yet to be firmly established, especially in the restaurant, front-of-the-house context.

SALES CONTESTS AS COMPENSATION

To investigate the appropriateness of SCs in the restaurant context using existing literature, we must use front-of-the-house restaurant servers as a proxy for salespeople. Most sales-related compensation is based on the idea that individual efforts should be sufficiently and fairly rewarded. In practice, though, outputs are easier to monitor than efforts, and determine the reward given. Such is the case in sales contests, one of the four types of sales force compensation (Basu, Srinivisan & Staelin, 1985).

According to previous research, the effectiveness of sales contests decreases as the variance in skill levels of the salesforce (front-of-the-house workers) increase (Lazear & Rosen, 1981). When employees competing in a sales contest are of varying skill levels, the differences in their perceived ability to win the contest may contaminate its effectiveness. This is usually the case in restaurant organizations, where employees may differ in the nature of their skills and the level of their skill development. In this case, adverse selection and moral hazard are likely to stand in the way of sales contest effectiveness (Holmstrom, 1979).

Adverse selection refers to the inability of a manager to determine that front-of-the-house employees with higher skill levels are in fact, achieving higher sales. While many managers may believe that they instinctually know which employees have higher skill levels directly through their achieved sales, the restaurant context makes this instinct questionable. The reason for this context's influence on instinct lies in the possibility that assumptions subsumed in many works on optimal sales contest design and formulation may not hold in restaurants. From this discussion, it is proposed that:

> P1a: *RSC effectiveness decreases as variance in participant skill levels increases.*
> P1b: *RSC effectiveness decreases as adverse selection increases.*

While context designs greatly vary, the assumptions of several studies limit generalizability to the restaurant context. For instance, many sales contests do not "account for sales territories with different potential" (Gonik, 1978, p. 116). Service territories for which restaurant employees are responsible may often be highly variable, whether structured or randomly assigned. Also, attempting to design an "optimal sales contest" (Kalra & Shi, 2001, p. 170) may be extremely sensitive to how sales are distributed over the course of the contest. In the restaurant industry, restaurant demand can fluctuate widely and sales can be difficult to forecast (for a review of studies, see Hu, Chen, & McCain, 2004; or Newberry, Klemz & Boshoff, 2003).

It is nearly impossible for a restaurateur to control all the aforementioned (and other unmentioned) variables that make established literature generalizable to this context. Part of this inability rests with the intangible and customer-specific range of skills needed to sell the restaurant experience; another rests with human resources issues such as turnover and training; yet another with server variability in encountering and handling different situations. These variables must be removed from the equation to know how skill drives sales, and to what degree.

All of these factors converge in the fact that both front-of-the-house employees and managers can never be certain about the exact sales that will result from a given effort. Initial experiments on uncertainty as a predictor variable show that in situations where the link between effort and outcome cannot be verified, incentives fail (Ghosh & John, 2000). In these cases, the relationship between effort and sales is not deterministic (Rao, 1990), which causes moral hazard. Moral hazard refers to the uncertainty employees may have that they will actually be rewarded for their higher efforts, which causes the sales contest to lose its incentive power. It is therefore proposed:

P2a: *RSC effectiveness decreases as the relationship between effort and outcome weakens.*
P2b: *RSC effectiveness decreases as moral hazard increases.*

Due to the above argument, the organization "needs to design a compensation scheme which recognizes the salespersons' different skills, and provides proper payments corresponding to those skill levels so that all salespersons have an incentive to achieve the sales most desirable from the firm's point of view" (Rao, 1990, p. 323). While a gratuity-based compensation scheme is able to do this, it is possible that sales contests may not.

SALES CONTESTS AS MOTIVATION

Two research gaps exist in the literature to inform SC application to both industry and theory. The first deals with the difficulties in clarifying and tying together motivation theories in the organizational behavior literature (Shamir, 1991). The fragmentation of motivation theories in general, a popular subtopic in SC studies, may have contaminated the application of these theories to SCs in many disciplines. Thus, the inability of motivation theories to be generalized across situations and cultures may also limit their ability to predict SC effectiveness in the restaurant context.

The second relates to tying together various perspectives–such as line-level employee, managerial, customer-oriented, and organizational–to yield a holistic model of sales contests that can be readily applied to restaurants. Linking various perspectives on SCs has yet to yield a strong holistic model that can be readily applied to restaurant. Prevalent streams of SC research–motivation and agency theories being the most popular foundations–have been garnered from samples that include service industries; but these samples differ in many ways from front-of-the-house restaurant workers.

Few to no scholarly studies have explicitly looked at restaurant sales contests (RSCs), leaving hospitality professionals with little context-specific grounded insight to this area of interest. Gaps in the literature have resulted in a relative lack of information and resources available to help restaurant professionals make profitable and strategically sound SC decisions in the future (for a selection of cited and reviewed studies, see Table 1; for more comprehensive reviews, see Kalra & Shi, 2001; Murphy & Dacin, 1998; Murphy, Dacin & Ford, 2004).

Sales contest studies have often, in the past, used motivation theories as their foundations (e.g., Kohn, 1993; Kurland, 1995). Motivation theories commonly used when investigating SCs have been scrutinized in recent years, though, due to assumptions and biases that may limit generalizability across industries and settings. For example, goal-setting SC theory (Locke and Latham, 1984), may suffer from an "overreliance on individualistic-hedonistic assumptions and...over-emphasis of cognitive-calculative processes" (Shamir, 1991, p. 405). The result of this imbalance is a situational bias of strength/weakness. Where the goals are clearly communicated, rewards are sufficient, and rewards are closely linked to performance, "situational strength" is high (Mischel, 1973). Thus, goal-setting "may be particularly useful in situations that strongly determine behavior and relatively useless in 'weak' situations where the variation in behavior is both larger and more reflective of variation in individual motivation" (Shamir, 1991, p. 407).

'Weak' organizational situations exist in the public sector (Perry & Porter, 1982), in more collectivist cultures (Hofstede, 1980), and in environments where "dispositional variables such as individual motivation are less likely to be relevant for the explanation of behaviour in those situations (Weiss, 1986)" (Shamir, 1991, p. 407). Organizations that use RSCs to motivate front-of-the-house employees may also suffer from low situational strength. Given the interdependent, complex, and multifaceted nature of the service process, the ability to strongly link rewards to performance is inherently problematic. The interdependence

TABLE 1. A Sample of Sales Contest Studies

Published Research	Method	Sample	Findings/Recommendations
Haring & Myers, 1954	Survey, descriptive statistics	Members of the National Sales Executives (n=396)	Respondents were all members of the National Sales Executives. 243 of 396 believed basic compensation was the #1 factor to secure above normal performance from the average salesman; sales contests #2
Wildt, Parker & Harris, 1980-1981	Non-empirical, 5 proposed steps for implementation	None	Call for further research, specifically recommending less reliance on case studies, input-output designs, managerial preferences, and assuming frequency of use indicates preference.
Wotruba & Schoel, 1983	Survey, descriptive statistics and correlations	Agents in 1 real estate company (n=16).	Many findings, including but not limited to: SCs resulted in higher profits when firms focused on better sales balance, lower profits when focused on increased sales volume or developing new sales skills. Negative side effects of SCs increased when firms sought to increase market share or higher-profit margin item sales
Luthans, Paul & Taylor, 1985	Field experiment	Departments in 1 retail store (n=16).	Process-based (rather than outcome-based) goals are positively related to frequency in SC-targeted salesperson behaviors.
Hart, Moncrief & Parasuraman, 1989; Moncrief, Hart & Robertson, 1988	Survey, descriptive statistics	Salespersons from 25 food brokerage (n=84), and from 1 cosmetics department (n=28).	SCs should end in advance of evaluation; aspects besides SC goals should be evaluated, especially cheating and damaged customer relations; SCs are difficult to discontinue once initiated; poor design is negatively related to morale; performance decline after SC end is typical
Murphy & Sohi, 1995	Survey, content analysis	Salespersons in a 100-member salesforce (n=28)	Individual differences of salespersons relate to their feelings regarding SCs. Career stage is negatively related to higher order rewards, as is age of the salesperson. Salesperson organizational commitment is positively related to SC satisfaction. Task-specific self esteem is positively related to the number of perceived SC advantages; and negatively related to the number of perceived SC disadvantages.
Beltramini & Evans, 1988	Survey, descriptive statistics and correlations	Direct salespersons from 3 companies (n=933).	A number of findings related to contest rewards, benefits, goals, budget, themes, timing and frequency, and outcomes. Respondents preferred simple, straightforward themes, frequently communicated peer progress, limited reach and participation potential of contests, and differences in SC participation based on career stage.

Published Research	Method	Sample	Findings/Recommendations
Murphy & Dacin, 1998	Non-empirical, 12 propositions	None	Salesperson perceptions of SC components–type and value of award, goal, and structure affect salesperson attitude, behavioral intent and behavior during SC.
Kalra & Shi, 2001	Non-empirical, 8 propositions	None	Agency theory used as foundation. Numerous recommendations on optimal contest design, depending on whether sales follow a uniform or logistic distribution; covers "only cases where the compensation structure is based on relative performance levels" (p. 171)
Radin & Predmore, 2002	Survey, t-tests	Buyers of consumer products: e.g., computers, audio-equipment, furniture, health & beauty aids, carpeting (n=417).	Measured non-expert consumer perceptions of purchase scenarios when consumers depend on the salesperson for information. Respondents indicated that ethical concerns of product-specific sales incentives did not significantly affect hypothetical purchase intentions or reliance on salesperson opinions
Murphy, Dacin & Ford, 2004	Survey, conjoint analysis	3 U.S. companies–consumer, industrial, and health-care sectors, 46 business units. n=796	Expectancy theory used as foundation. Respondents indicated preferences "for outcome-based goals, a limited number of winners (40%), durations of approximately a sales cycle (as appropriate for various industries)... higher award values (3 weeks' pay)... [and] cash awards" (p. 137)

of employees and customers in the process prevents isolating the causes of sales context efforts or outcomes, and the demand-driven nature of the restaurant business increases uncertainty in the organizational environment.

SALES CONTESTS AND TOTAL QUALITY MANAGEMENT (TQM)

Two decades ago, W. Edwards Deming made the prediction that total quality management would move into the service industries, including restaurants (1986). This prediction has materialized, with many restaurant corporations enacting Deming's 14 points of TQM in their operations (Kincaid, 2005). Ironically, many of these same corporations enacted periodic sales contests as well. Deming believed that motivating via rewards leads to an addiction of sorts. When one relies upon rewards to motivate behavior, the behavior is likely to last only as long as the reward is in place. If the contest is continued into the future, the reward will have to increase in size in order to have the desired effect. If management is unwilling to up the ante, the desired behavior will ultimately be extinguished. From this and previous evidence, the arguments suggest that:

P3: *The size of RSC rewards must increase as RSCs are repeated to maintain RSC effectiveness.*

Steve Kerr (1975), an organizational behavior scholar, would likely agree with Deming for reasons that are different but related. Kerr described a phenomenon regarding organizational tactics used in organizations. His thesis was that many reward systems encourage behavior *other* than that which management desires. A wine sales contest that rewards individuals for selling bottled wine may undermine management's larger goal of providing great service to all customers. How would a server respond to customers who do not wish to order wine? Might they be ignored in favor of those who do order wine? These questions and others point to the possibility that a sales contest could be an example of "The Folly of Rewarding 'A' While Hoping For 'B'" (Kerr, 1975).

P4: *RSC effectiveness decreases as the gap between customer-desired outcomes and RSC-rewarded outcomes increases.*

MENU AND FINANCE IN RSC OBJECTIVES

Restaurant operators often focus on check average or particular menu items/categories as a means of increasing total revenue. This view is narrow and shortsighted given recent applications of revenue management concept to restaurants (Kimes & Barrash, 1999). For example, most restaurant patrons can reasonably be expected to order a main course. In terms of food, the add-ons that increase check average are appetizers and desserts–common RSC targets. Selling desserts may reduce total revenue and profit, however, by extending service cycle time and decreasing table turns. This makes RSCs such as "Menu Bingo" (Sullivan, 2002), a contest that rewards servers for selling one item in each menu category, sometimes counterproductive and detrimental to actual RSC objectives. Such RSCs provide evidence of Kerr's thesis in a restaurant context–managers often reward behaviors in opposition to organizational objectives.

Restaurant managers must be clear about the organization's goals, and must communicate them clearly to servers. In most cases, the most beneficial long- and short-term organizational objectives are to maximize revenue, profit and service/product quality concurrently. If it is possible for a sales contest to concurrently achieve these three objectives, the contest should cease to become a "bonus" or "special effort" rather, it should be built into the permanent system of the restaurant operation. Deming (1986; 2000) explains that management's responsibility for design and continuous improvement of the system of production makes all three of these objectives feasible.

Continuous training from day one, Deming states (1986) is the key to producing quality, revenue and profit. This assertion may explain the success of currently reported restaurant contests that reward for skills and training (Cebrynski, 2002). Contests that reward for sales, however, shift the focus away from service and providing a high-quality dining experience.

Also, achieving training objectives is under the server's control, and expressing well-trained behavior is an outcome that is more aligned with the true effort of workers throughout a restaurant operation. In order to increase wine sales, for instance, educating servers on wine service and product attributes may prove more productive and goal-oriented than simply awarding the top seller of wine in a given time period. An RSC, in this instance, rewards an outcome that is not directly connected to effort and alienates all but the most highly trained sellers–note, not the best servers. Therefore:

> P5a: *Rewards for skill have a higher positive correlation with product quality than rewards for sales.*
> P5a: *Rewards for skill have a higher positive correlation with long-term revenue than rewards for sales.*
> P5a: *Rewards for skill have a higher positive correlation with long-term profit than rewards for sales.*

COMMON RSC TROUBLE SIGNS

Seminal studies on sales contests and related incentives outline common SC trouble signs, all rooting in Deming's TQM principles (for a comprehensive outline, see Orsini, 1987). These trouble signs are side effects of the motivational strength of RSCs, the first observable ones being:

- Preoccupation with meeting goals
- Failure to carry out normal routines

- Decrease in effective intracompany communication
- Increase in customer dissatisfaction

These phenomena are the result of strong goal orientations. The job definition has changed; management has made clear where time and energy is to be spent. Work or problems that do not relate to the specified goals take a back seat to the new thrust ... Interestingly, management is often fooled into believing that the incentive bonus plan is producing the desired results when it is not, since employees will work very cleverly to please management and thereby collect the bonus (Orsini, 1987, p. 182).

It is not hard to imagine these negative results occurring in a restaurant context. Suppose a restaurant operator wishes to increase total revenues. This is essentially a goal shared with the prospective RSC participants, tipped employees who typically earn more when total revenues increase. Rewarding this goal would be redundant, opening the door of increased competition and decreased cooperation–however the contest defines the boundaries of participation–for few to no apparent reasons.

Making higher profit items an RSC target might have an improved effect on RSC outcomes, but at the possible detriment to customers. In a demand-driven economy where customers want what they want when they want it, overly suggestive selling may be seen as intrusive, unwanted, or even suspect (e.g., Radin & Predmore, 2002). Additionally, many RSCs could not, in practice, guarantee against subversive employee behavior, or cheating, in all cases. One example of this might be ringing up a non-RSC-rewarded item as one that would serve self-interests and win the RSC prize.

Admittedly, these are sizable disadvantages. According to Deming, however, one of the largest problems is the inability of RSCs to factor in sources of variation beyond the participant's control. The server that is assigned a table of eight with four adults and four children may be hard-pressed in convincing the group to purchase several hard liquor drinks; while a party stopping in for coffee and dessert may be worthless in helping a server up-sell the appetizer menu. And for RSC participants across several job positions, regular customers served often settle into a product/service pattern that can threaten gratuities and/or loyalty if disrupted. There are many cases in which RSCs can harm the organization, both in the short- and the long-term.

RSCs AND UNETHICAL SALES BEHAVIOR

As motivation is the primary rationale underpinning sales contest use (e.g., Beltramini & Evans, 1988), restaurant operators may also consider the impact RSCs have on the ethical nature of participants' behavior. If RSCs truly motivate, what do they motivate participants to do? In the restaurant context, successful managers are characterized by honesty, integrity, and strong ethics (Reynolds, 2000), as setting these standards is vital to service-based organizational cultures; but it is not clear that line-level employees share these characteristics. While unethical sales behavior has not been empirically targeted within the context of RSCs, research in other contexts raises concerns about its possible existence.

Front-of-the-house restaurant staff–notably servers and bartenders–are usually the most likely to participate in RSCs, and also earn the majority of their compensation through gratuities. As commission-based forms of compensation, both gratuities and RSCs are significant extrinsic motivators for RSC participants–but what happens when employees try to maximize both? Does anything have to "give?" While the answer to this question lays in the contest goals, compensation is a significant extrinsic motivator and behavioral driver in workplaces across many contexts. In the foundation psychological constructs involved, restaurants are no different.

When emphasis is placed upon short-term sales performance, reinforced through a commission-based compensation structure, an atmosphere is created that is ripe for ethical abuse (Bellizzi & Hite, 1989). Method of compensation (salary-based versus commission-based) has been found to influence ethical behavior in other organizational contexts (e.g., Kurland, 1991; 1995; 1996), and income level to negatively predict perceived employee opportunities for unethical behavior (e.g., Honeybutt, Glassman, Zuguelder, & Karande, 2001; Laczniak & Murphy, 1993). Quite simply, unethical behavior occurs more often when it is rewarded (e.g., Hegarty & Sims, 1979). In the aforementioned study, when unethical behavior resulted in greater profits for the organization, the unethical behavior occurred more frequently.

Income has also been found in non-restaurant contexts to be negatively related to perceived ethical problems in the workplace and more optimism about ethical behavior's link with personal success (Chonko & Hunt, 1985; Finn, Chonko, & Hunt, 1988). The RSC dilemma over the degree to which restaurant employees' income level and compensation method affect both the opportunities for and engagement in unethical behaviors has received exploratory qualitative support (Kincaid,

2004). These preliminary results show that RSCs may be counterproductive by reinforcing self-serving interests of employees rather than those of the customer and employer. Based on these discussions:

> P6: *If unethical behavior is perceived to increase the probability of an RSC reward more than ethical behavior, the incidence of unethical behavior will increase.*

RECOMMENDATIONS FOR FUTURE RESEARCH

The authors recommend that the use of contests, with sales-related and other objectives, should be empirically investigated in the restaurant setting, examining both short- and long-term objectives, so that the issues discussed herein can be tracked and clearly evaluated. While the existing literature is helpful in developing conceptual frameworks for study, the assumptions, industry characteristics, and nature of the variables involved may diverge significantly. A restaurant-specific stream of research may shed light on both theory and practice in this area.

As organizations that depend on employee efforts for contest effectiveness, such as restaurants, may be better served by exploring the employee perspective on these incentives (Murphy & Dacin, 1998). As noted by Murphy and Dacin (1998), inclusion of employee feelings, attitudes and motivations has often been disregarded in favor of input-output based foci (Anderson, Crowell, Success, Gilligan & Wikoff, 1983; Caballero, 1988; Luthans, Paul & Baker, 1981; Luthans, Paul & Taylor, 1985; Wildt, Parker & Harris, 1987; Wotruba & Schoel, 1983). Future research might, for example, compare the relative impacts of RSC participant and RSC designer perspectives on related organizational outcome variables.

In addition, foci from financial (e.g., total revenue, overall profit margin) to marketing (e.g., customer satisfaction, brand management), and from motivation (e.g., employee attitudes, feelings, and behaviors) to compensation (e.g., reward valence, equity, and allocation) may clarify RSC impacts on particular stakeholders in restaurant operations, encouraging more cooperation in formulating strategy. For instance, a simply-designed survey paired with secondary data may suggest that although one group of stakeholders perceives great benefits stemming from RSCs, another does not. Those perceptions could then be compared to trends in product quality, revenues and profits from a short- and long-term perspective. Again, controlling for the many variables inherent in restaurants may limit reliability, perhaps calling for experimental approaches. At such an early stage of development, both qualitative and quantitative inquiries would serve to advance knowledge of RSC phenomena.

CONCLUSION

Drawing on work in the SC literature, both scholarly in the general business discipline and anecdotal in the restaurant-specific discipline, a wide range of RSC issues were explored here including: motivation as underlying theory; menu, finance, employee behaviors, and ethics as outcomes. While the business disciplines generally agree that there is theoretically an optimal sales contest structure and content, dissention continues over the negative impact of SCs in different contexts. We contend here that restaurants, as (1) a service-based and customer-centered context where (2) revenue variation is often out of RSC participant control and (3) participants are already motivated to achieve many possible RSC goals, are indeed a context where SCs may prove to do more harm than good. Especially in light of recent findings showing ethical behavior as a behavioral characteristic of successful restaurant managers (Reynolds, 2000), RSCs should be approached with caution.

Additional theoretical and empirical work is needed to clarify, isolate, and test the many causes and effects inherent in the complex restaurant environment. Attributing sales volume to the presence or absence of a sales contest may prove to be too simplistic, masking or sidestepping system-wide problems that Deming's TQM model might solve. Common to the alternatives is that they are harder work for management than are RSCs. To guide future empirical work in the SC area, particularly in the restaurant context, we devised a set of testable propositions that flow from the literature presented herein.

A sales contest is an easily administered, quick fix approach. However, like most quick fixes, a sales contest fix has not proven effective in the long run. Although they may be intuitively appealing, RSCs may be ultimately self-defeating.

REFERENCES

Anderson, D. C., Crowell, C. R., Sucee, J., Gilligan, K. D., & Wikoff, M. (1983). Behavior management of client contacts in a real estate brokerage: getting agents to sell more. *OBM in Multiple Business Environments*. Binghamton, NY: The Haworth Press, pp. 67-95.

Bartol, K. M. (1999). Reframing Salesforce Compensation Systems: An Agency Theory-Based Performance Management Perspective. *Journal of Personal Selling & Sales Management 19*(3), 1-16.

Basu, A. K., Lal, R., Srinivasan, V., & Staelin, R. (1985). Salesforce Compensation Plans: An Agency Theoretic Perspective. *Marketing Science 4*(4), 267-291.

Bellizzi, J., & Hite, R. (1989). Supervising unethical sales force behavior. *Journal of Marketing 53*(2), 36-47.

Beltramini, R. F., & Evans, K. R. (1988). Salesperson Motivation to Perform and Job Satisfaction: A Sales Contest Participant Perspective. *Journal of Personnel & Sales Management* 8(3), 35-42.

Berta, D. (2004). Restaurant companies boost employee morale with contests, games. *Nation's Restaurant News 38*(26), 6-7.

Boyes, W. J., Mounts Jr., W. S., & Sowell, C. (2004). Restaurant Tipping: Free-Riding, Social Acceptance, and Gender Differences. *Journal of Applied Social Psychology 34*(12), 2616-2628.

Caballero, M. J. (1988). A Comparative Study of Incentives in a Sales Force Contest. *Journal of Personal Selling & Sales Management 8*(1), 55-58.

Cebrzynski, G. (2002). Huddle House drops promo, repeats incentive contest. *Nation's Restaurant News 36*(37), 14.

Churchill, G. A., Ford, N. M., & Walker, O. C. (1993). *Sales Force Management.* Homewood, IL: Irwin.

Chonko, L., & Hunt, S. (1985). Ethics and marketing management: An empirical examination. *Journal of Business Research 13*(4), 339-359.

Coughlan, A. T., & Sen, S. K. (1989). Salesforce Compensation: Theory and Managerial Implications. *Marketing Science 8*(4), 324-342.

Crusco, A. H., & Wetzel, C. G. (1984). The Midas touch: The effects of interpersonal touch on restaurant tipping. *Personality and Social Psychology Bulletin 10*, 512-517.

Deming, W. E. (1986). *Out of the crisis.* Cambridge, MA: Massachusetts Institute of Technology, Center for Advanced Engineering Study.

Deming, W. E. (2000). *The new economics for industry, government, and education.* Cambridge, MA: MIT Press.

Farley, J. U. (1964). An Optimal Plan for Salesmen's Compensation. *Journal of Marketing Research 1*(2), 39-43.

Finn, D., Chonko, L., & Hunt, S. (1988). Ethical problems in public accounting: The view from the top. *Journal of Business Ethics 7*(8), 605-615.

Garrity, K., & Degelman, D. (1990). Effect of server introduction on restaurant tipping. *Journal of Applied Social Psychology 20*, 168-172.

Ghosh, M., & John, G. (2000). Experimental Evidence for Agency Models of Salesforce Compensation. *Marketing Science 19*(4), 348-365.

Gonik, J. (1978). Tie salesmen's bonuses to their forecasts. *Harvard Business Review 56*(3), 116-122.

Green, J. R., & Stokey, N. L. (1983). A comparison of tournaments and contracts. *Journal of Political Economics 91*(3), 349-364.

Haring, A., & Myers, R. H. (1953). Special Incentives for Salesmen. *Journal of Marketing 18*(2), 155-159.

Harris, M., & Raviv, A. (1979). Optimal Incentive Contracts with Imperfect Information. *Journal of Economic Theory 20*(2), 231-259.

Hart, S. H. (1985). An Empirical Investigation of Salespeople's Behavior, Effort, and Performance During Sales Contests. *Dissertation Abstracts International 45*(9-A), 2932.

Hart, S. H., Moncrief, W. C., & Parasuraman, A. (1989). An Empirical Investigation of Salespeople's Performance, Effort and Selling Method During A Sales Contest. *Journal of the Academy of Marketing Science 17*(1), 29-39.

Hegarty, W., & Sims, H. (1979). Organizational philosophy, policies, and objectives related to unethical decision behavior. A laboratory experiment. *Journal of Applied Psychology 64*(3), 331-338.

Hofstede, G. (1980). *Culture's consequences: international differences in work-related values.* Beverly Hills, CA: Sage.

Holmstrom, B. (1979). Moral Hazard and Observability. *Bell Journal of Economics 10*(1), 74-91.

Honeycutt, E., Glassman, M., Zuguelder, M., & Karande, K. (2001). Determinants of ethical behavior. *Journal of Business Ethics 32*(1), 69-79.

Hu, C., Chen, M., & McCain, S. C. (2004). Forecasting in Short-Term Planning and Management for a Casino Buffet Restaurant. *Journal of Travel & Tourism Marketing 16*(2/3), 79-98.

John, G., & Weitz, B. (1989). Salesforce Compensation: An Empirical Investigation of Factors Related to Use of Salary Versus Incentive Compensation. *Journal of Marketing Research 26*(1), 1-14.

Joseph, K., & Kalwani, M. U. (1998). The Role of Bonus Pay in Salesforce Compensation Plans. *Industrial Marketing Management 27*, 147-159.

Kalra, A., & Shi, M. (2001). Designing Optimal Sales Contests: A Theoretical Perspective. *Marketing Science 20*(2), 170-193.

Kerr, S. (1975). On the folly of rewarding "A" while hoping for "B." *Academy of Management Journal 18*, 769-783.

Kimes, S. E., & Barrash, D. I. (1999). Developing a restaurant management strategy. *Cornell Hotel & Restaurant Administration Quarterly 40*(5), 18-30.

Kincaid, C. (2005). Personal interview. July 16. Las Vegas, NV.

Kincaid, C. (2004). An examination of the effect of ethical climate on ethical optimism and organizational commitment. *Dissertation Abstracts International 64*(12), 4523. (UMI No. 3115892).

Kohn, A. (1993). Why Incentive Plans Cannot Work. *Harvard Business Review 71*(5), 54-63.

Kotler, P. (1997). *Marketing Management: Analysis, Planning, Implementation and Control*, 9th ed. New York: Prentice Hall.

Kurland, N. B. (1991). The Ethical Implications of the Straight-Commission Compensation System: An Agency Perspective. *Journal of Business Ethics 10*(10), 757-766.

Kurland, N. B. (1996). Sales Agents and Clients: Ethics, Incentives, and Modified Theory of Planned Behavior. *Human Relations 49*(1), 51-74.

Kurland, N. B. (1995). Ethics, Incentives, and Conflicts of Interest: A Practical Solution. *Journal of Business Ethics 14*(6), 465-475.

Laczniak, G. & Murphy, P. (1993). Incorporating marketing ethics into the organization. *Marketing Ethics: Guidelines for Managers.* Lexington, MA: D. C. Heath.

Lal, R. & Staelin, R. (1986). Salesforce Compensation Plans in Environments with Asymmetric Information. *Marketing Science 5*(2), 179-198.

Lazear, E. P., & Rosen, S. (1981). Rank-order tournaments as optimum labor contracts. *Journal of Political Economics 89*, 843-864.

Luthans, F., Paul, R. & Baker, D. (1985). An Experimental Analysis of the Impact of Contingent Reinforcement on Salespersons' Performance Behavior. *Journal of Applied Psychology 66*, 314-323.

Luthans, F., Paul, R., & Taylor, L. (1985). The Impact of Contingent Reinforcement on Retail Salespersons' Performance Behaviors: A Replicated Field Experiment. *Journal of Organizational Behavior Management 7*(1/2), 25-35.

Lynn, M., & Mynier, K. (1993). Effect of server posture on tipping. *Journal of Applied Social Psychology 23*, 678-685.

Marchetti, M. (2004). Why Sales Contests Don't Work. *Sales & Marketing Management 156*(1), 19.

Marvin, B. (1997). *Guest-Based Marketing: How to Increase Restaurant Sales Without Breaking Your Budget.* New York: John Wiley & Sons, Inc.

Mischel, W. (1973).Toward a cognitive social learning reconceptualization of personality. *Psychological Review 80*, 200-213.

Mok, C. & Hansen, S. A. (1999). Study Of Factors Affecting Tip Size In Restaurants. *Journal of Restaurant & Foodservice Marketing 3*(3/4), 49-64.

Moncrief, W. C., Hart, S. H., & Robertson, D. (1988). Sales Contests: A New Look at an Old Management Tool. *Journal of Personal Selling & Sales Management 8*(3), 55-61.

Murphy, W. H. & Dacin, P. A. (1998). Sales Contests: A Research Agenda. *Journal of Personal Selling & Sales Management 18*(1), 1-16.

Murphy, W. H., Dacin, P. A. & Ford, N. M. (2004). Sales Contest Effectiveness: An Examination of Sales Contest Design Preferences of Field Sales Forces. *Journal of the Academy of Marketing Science 32*(2), 127-143.

Murphy, W. H. & Sohi, R. S. (1995). Salespersons' perceptions about sales contests. *European Journal of Marketing 29*(13), 42-65.

Nalebuff, B. & Stiglitz, J. E. (1983). Prizes and incentives: Toward a general theory of compensation and competition. *Bell Journal of Economics 14*, 21-43.

Newberry, C. R., Klemz, B. R., & Boshoff, C. (2003). Managerial implications of predicting purchase behavior from purchase intentions: a retail patronage case study. *Journal of Services Marketing 17*(6/7), 609-620.

Orsini, J. N. (1987). Bonuses: What is the Impact? *National Productivity Review (Spring)*, 180-184.

Perry, J. L., & Porter, L. W. (1982). Factors affecting the context of motivation in public organizations. *Academy of Management Review 7*, 83-98.

Pratt, J. W. (1964). Risk Aversion in the Small and in the Large. *Econometrica 32*(1), 122-136.

Radin, T. J. & Predmore, C. E. (2002). The Myth of the Salesperson: Intended and Unintended Consequences of Product-Specific Sales Incentives. *Journal of Business Ethics 36*, 79-92.

Rao, R. C. (1990). Compensating Heterogeneous Salesforces: Some Explicit Solutions. *Marketing Science 9*(4), 319-341.

Reynolds, D. (2000). An Exploratory Investigation Into Behaviorally Based Success Characteristics of Restaurant Managers. *Journal of Hospitality & Tourism Research 24*(1), 92-103.

Ross, S. (1973). The Economic Theory of Agency: The Principal's Problem. *American Economic Review 63*(2), 134-139.

Shamir, B. (1991). Meaning, Self and Motivation in Organizations. *Organization Studies 12*(3), 405-425.

Shavell, S. (1979). Risk Sharing and Incentives in the Principal and Agent Relationship. *Bell Journal of Economics 10*(1), 55-73.

Sullivan, J. (2002). Sales contest basics inspire competition and profits. *Nation's Restaurant News 36*(35), 16-17.

Wildt, A. R., Parker, J. D. & Harris, C. E. (1980-1981). Sales Contest: What We Know and What We Need To Know. *Journal of Personal Selling and Sales Management 1*(3), 57-62.

Wotruba, T. R. & Schoel, D. J. (1983). Evaluation of Salesforce Contest Performance. *The Journal of Personal Selling & Sales Management 3*(2), 1-10.

doi:10.1300/J369v09n02_06

A Recipe for Success: The Blending of Two Disparate Nonprofit Cultures into a Successful Collaboration– A Case Study

Jane E. Barnes
Susan G. Fisher

SUMMARY. The number of nonprofit organizations continues to increase and, like their for-profit counterparts, they face issues of strategy, governance and growth. Collaboration is increasingly becoming one way for the NPO to remain competitive and provide better client services in the face of competition from other agencies vying for dwindling government funds. Cultural clashes can often be the downfall of an alliance. However, because employees of nonprofits tend to be more dedicated to their mission, even disparate cultures can collaborate successfully through hard work, open communication, and focus on the ultimate goal.

Jane E. Barnes, PhD, is Assistant Professor, in the School of Business, at Meredith College, 233 Harris Hall, Raleigh, NC 27607-5298 (E-mail: BarnesJ@meredith.edu).

Susan G. Fisher, PhD, RD, LD, is Associate Professor, in the Department of Human Environmental Sciences, at Meredith College, 224 Martin Hall, Raleigh, NC 27607-5298 (E-mail: fishers@meredith.edu).

[Haworth co-indexing entry note]: "A Recipe for Success: The Blending of Two Disparate Nonprofit Cultures into a Successful Collaboration–A Case Study." Barnes, Jane E., and Susan G. Fisher. Co-published simultaneously in *Journal of Foodservice Business Research* (The Haworth Hospitality & Tourism Press, an imprint of The Haworth Press, Inc.) Vol. 9, No. 2/3, 2006, pp. 111-125; and: *Human Resources in the Foodservice Industry: Organizational Behavior Management Approaches* (ed: Dennis Reynolds, and Karthik Namasivayam) The Haworth Hospitality & Tourism Press, an imprint of The Haworth Press, 2006, pp. 111-125. Single or multiple copies of this article are available for a fee from The Haworth Document Delivery Service [1-800-HAWORTH. 9:00 a.m. - 5:00 p.m. (EST). E-mail address: docdelivery@haworthpress.com].

Available online at http://jfbr.haworthpress.com

doi:10.1300/J369v09n02_07

Using a case study involving two foodservice firms, this paper examines the role that culture can play in a nonprofit alliance and offers practical suggestions for administrators contemplating collaboration with another NPO. doi:10.1300/J369v09n02_07 *[Article copies available for a fee from The Haworth Document Delivery Service: 1-800-HAWORTH. E-mail address: <docdelivery@haworthpress.com> Website: <http://www. HaworthPress.com>*

KEYWORDS. Nonprofit, strategy, governance, growth

INTRODUCTION

This paper addresses the role that culture can play in a nonprofit (NPO) alliance through a case analysis of the collaboration of two foodservice firms. The number of nonprofit agencies has grown over the past decade (Giffords & Dina, 2004). Like their for-profit counterparts, nonprofit organizations face issues of strategy, governance and growth. As their client bases increase, they look for other ways to serve the community than just adding staff or volunteers. Forming alliances, collaborations or mergers is becoming necessary for the NPO to remain competitive and provide better client services (Giffords & Dina, 2003). Skloot (2000) has even raised the question as to whether NPOs as they are currently structured can exist in today's marketplace.

In the current environment, nonprofits have been pressured to do more with less and to adopt more business-like strategies (Hiland, 2003). Government support to resolve social problems is being pushed back to local agencies that are competing with other nonprofits for community support and funding (Gifford & Dina, 2003). At the same time, they are competing with other organizations for declining government funds (Boardman & Vining, 2000). To assure the survival of their organization, or to resolve a strategic financial issue, NPOs are increasingly looking toward a combination with another nonprofit firm. When the decision to take this step is made, NPOs have several options: mergers, acquisitions or alliances. While the first two are increasingly used to increase organizational effectiveness or to relieve financial pressures (La Piana & Hayes, 2005), the latter is intended to help nonprofits become more flexible, competitive and innovative–they are essentially to NPOs what mergers and acquisitions are to for-profits (Bates, 1999). Forming an alliance between the agencies may provide many benefits including: achieving a stronger position in the community, increasing the response to community needs, or boosting professional recognition (McLaughlin, 1996).

THE "SOFT" FACTORS IN A COLLABORATION

Although the literature on organizational combinations is replete with merger and acquisition cases in the for-profit sector, there is much less information available on nonprofit mergers and even less on NPO alliances. Furthermore, most of the outcome studies have examined "hard" organizational factors such as product integration and staffing, but only recently have the studies focused on the importance of "softer" factors such as organizational culture or other elements involving a "human" touch (Cartwright & Cooper, 1995). In a nonprofit alliance in which the organizations tend to be smaller and the parties are often passionate about their work, issues surrounding governance, integration and implementation can become quite personal, and success is often determined by the personalities involved.

The following case study is used to demonstrate the importance of these softer factors on successful alliances, and will discuss the role that culture plays in governance, integration and, ultimately, the success of the collaboration. This paper will also identify practical lessons for those nonprofit administrators contemplating an alliance.

The case method has been commonly used to study collaborations between nonprofit agencies (see, for example, Giffords & Dina, 2003, 2004; O'Brien & Collier, 1991). In this study, one of the authors was directly involved in the collaboration through her role as a board member of one of the organizations. Her experiences, interviews with key members of both organizations, past and present, and organizational records (meeting minutes, bylaws, etc.) formed the basis of the data gathering. Personal interviews with the three current directors and several former board members were conducted by one author who also made personal visits to the facilities. Transcriptions by this author were provided to the interviewees for their review; drafts of the paper were also sent to the directors for their concurrence.

THE CASE STUDY

The Collaboration: Meals on Wheels + The Food Shuttle = Food Runners

Food Runners Collaborative, Inc. was formed in 1999 when two nonprofit organizations with missions to address hunger decided to collaborate on the construction of a building that would serve both their needs.

Meals on Wheels of Wake County (MOW), a private non-profit that receives government funding, was concerned about being tied to a single source food provider and, with no other competing bidders, proposed building its own kitchen facility. The Inter-Faith Food Shuttle (Food Shuttle or IFFS), formed as a grass roots movement in 1989, needed refrigeration and a loading dock, and desired a kitchen facility for its culinary school.

Meals on Wheels. Meals on Wheels of Wake County, NC, was founded in 1974 as a 501 (c)(3) nonprofit organization. It has over 2,200 volunteers and serves 1300-1400 meals per day to homebound elderly adults, people with disabilities, and to eight Senior Nutrition Centers (congregate sites) in the greater Raleigh area. Its mission is to provide nutritious meals to the homebound and elderly and its standards are set by the federal and state governments. There is a waiting list of approximately 700 people to get on one of their home-delivery routes.

MOW receives funding from The United Way, government grants, fundraisers and donations, among other sources. Federal funding dictates that meals must meet one-third of the daily recommended intakes for the elderly. Concerned that, for several years, only one food provider submitted bids to provide meals, the MOW board of directors proposed building its own kitchen facility to have control over its meal supply system. They hired a fund raising consultant and were close to signing a contract with a development firm to do the fundraising for a new building when they learned of the Food Shuttle's need for kitchen facilities.

Inter-Faith Food Shuttle. IFFS was chartered in 1989 to provide safe, sanitary distribution of surplus perishable food to those in need. Beginning as an all-volunteer grassroots movement, the Food Shuttle collected donated food in coolers and delivered it to local soup kitchens, the Salvation Army and other area agencies. In 1992, they became a 501(c)(3) nonprofit organization, bought a refrigerated truck, and moved into their first building; later, they began an effort to combat hunger at its root cause by establishing the Culinary Job Training Program which trains homeless and sheltered people in basic culinary skills, preparing them for reemployment in the foodservice industry. The Food Shuttle currently serves 200 nonprofit agencies in a seven county area. Its clients include shelters, missions, community organizations, outreach pantries, soup kitchens and senior centers. Most of the food is donated from the local Farmers' Market and food retailers.

In 1999, Food Shuttle had grown to the point that they could not handle all of the food they were collecting. Because of the lack of adequate refrigeration and a loading dock, more than fifty percent of the donated food was passed on without gleaning[1]. The Food Shuttle had need for a blast freezer and adequate cooling to preserve food and reduce spoilage in order to provide more usable food to feed more people. In addition, the culinary program was growing and they were looking for a better-equipped kitchen facility. All this prompted the IFFS board to draw up proposals to build their own kitchen facility and hire counsel to start a fund raising campaign.

Food Runners Collaborative, Inc. Through a mutual friend who encouraged them to work together, the directors of both organizations met in early 1999 to discuss their mutual goal: a kitchen facility to serve the community. Although MOW was close to signing a contract for a capital campaign, the board believed that due diligence required a discussion with the Food Shuttle. At that time, IFFS was further along in their planning for a building, while MOW had more fund raising plans. Four members from each organization met over the course of three to four months to consider the collaboration. The first meeting was full of mutual optimism and excitement. Eventually, the mutual desire to feed the community and the potential problems of two similar agencies raising money for basically the same goal outweighed any concerns they may have had over the collaboration.

In the next few months, however, it was determined that a separate third organization would need to be formed in order to raise money for the facility. Both boards approved the name Food Runners Collaborative (FRC), which was incorporated in late 1999. In March 2000, the first bylaws established a board comprised of seven total board members, although in the articles of incorporation only six were listed–three from each organization.

THE EFFECTS OF CULTURE ON COLLABORATIONS

Organizational culture is the shared values and beliefs that underlie a company's identity (Schein, 1992). It is derived from the attitudes, behaviors, work rules and customs found in the workplace. According to Edgar Schein, a leading organizational development expert, organizational culture can be analyzed along three dimensions: artifacts, which describe structures and processes; espoused values, which include strategies and philosophies; and the basic underlying assumptions that form the unconscious, taken-for-granted beliefs and perceptions of the organization (1992:17). History, ownership, size, technology, leadership and the external environment all affect culture (Arsenault, 1998). There is no time in an organization's life when culture is more important than when it is faced with the prospect of consolidation (Arsenault, 1998).

NPOs usually have value-driven staffs, that is, the employees have chosen to do this particular work because of very powerful personal value-based choices (Arsenault, 1998). Every person in a nonprofit is dedicated to its mission, and what carries a mission is the organization's values–the implicit statements about what matters to the people associated with the organization (McLaughlin, 1998). Values are what underlie the behavior in an organization (McLaughlin, 1998). Over time, the nonprofit's leadership will direct the organization to be true to its underlying values.

The glue that links the values is the organization's culture, and this is where the greatest divergence among nonprofits is found (McLaughlin, 1998). For example, there are likely to be many people in a given area who simultaneously see the need for a specific service such as helping the homeless. It is also probable that their values are similar. Where they may differ, however, is in the ways that they make those values work–the culture that develops to link values to action (McLaughlin, 1998). These cultural differences are expressed in the visible artifacts (e.g., physical space, language, stories) and the unwritten assumptions (unconscious philosophies and moral codes).

Organizational culture has been found to be a key determinant in the success or failure of an organization (Flannery & Deiglmeier, 1999). Nonprofit collaborations, like their for-profit counterparts, often fail due to culture clashes and incompatibility (Buono & Bowditch, 1989; Hiland, 2003). In addition, basic differences in employees' behaviors, thoughts or actions may harm the implementation of the collaboration's goals (Olie, 1994).

Differing organizational histories. One of the key components to establishing an organization's culture is its history, including stories, legends and myths surrounding its founding (Shein, 1992). In this case, Meals on Wheels was affiliated with a national organization of the same name (although it was independently operated), received funding from the government, and had been in operation over twice as long as the Food Shuttle. MOW ran like a business: it had specific customers on regular routes with nutritional guidelines to follow. Because of the government standards around nutrition, it was regularly audited. MOW had a strategic plan, fund raising strategies and target goals. Its mission was to provide nutritional meals to the homebound and elderly.

The Food Shuttle culture derived from a much different history. The Inter-Faith Food Shuttle was formed when two friends–one Christian and one Jewish–lamented the waste of perfectly good food being thrown away while dining at a fast food restaurant. Initially, there were few formal principles other than getting as much donated food to as many hungry people as safely as possible[2], and there were no nutritional guidelines to follow. If MOW was run more as a business, IFFS operated as if they were all part of one big family. They saw their mission as salvaging usable food and putting it into the hunger community.

Combining two organizations with such different histories is not easy (O'Brien & Collier, 1991). Cultural issues often occur when the backgrounds of the organizations are different, for example, between a grass roots organization and one such as an arts council in which most of the group has mission-specific education. In a combination, those individuals who identify with this "class-related" aspect of the service may have a hard time accepting a partner who does not share the same characteristics (McLaughlin, 1998).

In this case study, it took a while to work toward common communication and mutual respect. At first, differences in funding sources and operations were sources of misunderstanding. The operating styles of the two parties, while both effective, were very different. MOW was comfortable in its established ways; it knew the size of its "sand box," was highly functional, and had a formal business planning process. MOW was limited by its defined, narrow mission and also by the size of its facilities, which did not allow for much expansion. Meals on Wheels saw itself as one large, extended family, with volunteers all over the county.

For the Food Shuttle, being one big family was one of its highest values. They intentionally got as many diverse people as possible into the family and used the family concept to shape their mission. The Food Shuttle saw its sand box as the beach and was more visionary, almost entrepreneurial. They viewed business planning as a creative, developmental process, one that was non-linear, and took advantage of opportunities as they arose rather than sticking rigidly to a plan.

Since the collaboration, their cultures have remained distinct, yet each organization has changed for the better. MOW, while still business-like, is thinking more out of the box on solutions to feed hungry seniors. The Food Shuttle is still visionary, but has become more operational and organized. The two parties can be described as elements of the kitchen of a successful restaurant: The best restaurants need both a head chef and a pastry chef–two talented individuals with different skills that complement one another and work together to make the restaurant a success.

Culture and Governance

Governance really matters to the future mission and effectiveness of the collaboration (McLaughlin, 1998). Because NPOs are mission-focused (rather than focused on the bottom line), they attract board members who are motivated by a passion for the organization's cause (La Piana & Hayes, 2005). The nonprofit board is responsible for accounting to its many stakeholders such as its funders and the public at large (McClusky, 2002). Nonprofit boards have a fiduclary responsibility to the organization, that is, they must act as if they own it, even though they do not. In his article on nonprofit governance, McClusky sets out eleven functions of NPO boards including determining the organization's mission and purpose, raising money, strategic planning, and organizing itself so that the board operates effectively (2002: 542).

Board composition. Governance is an area of uncertainty in a collaboration because there are no "recipe cards" to follow when combining two boards of directors or creating a new board (McLaughlin, 1998). Rules and processes exist for the board of merged organizations, but there is very little guidance for collaborations. In general, formulas for board membership tend to look backward (what was the composition before collaboration?) rather than forward (what is the vision of the new organization?). There is not any automatic connection between what any individual board member can offer and the organization from which he or she came (McLaughlin, 1998). Instead, organizations should staff the board with members who are key to the mission and who have skills necessary for the future. Furthermore, to be successful, it is suggested that the new board should sever the implied connection between individuals and the nonprofit organization with which they were previously associated (McLaughlin, 1998).

In this case study, the original articles of incorporation called for seven members to be on the Food Runners' board, and three from each organization were named. Soon after, the board was expanded from the original seven members to seven members each nominated from the two organizations. The details of how Food Runners would run had not been worked through by the board of directors, which was more concerned with building the kitchen facility than financial accountability. This is not unusual. Boards often have difficulty setting an overarching vision within which the individual missions make sense (Arsenault, 1998), or are concerned that the mission of their organization may change or their identity may be lost (McLaughlin, 1998). As a result, when the Food Runners' board initially met, there were problems with decision making and how the organizations related to each other. Not as much was accomplished as could have been in part because members weren't sure how open to be with one another. Again, this is not atypical. Board members are often attached to the status quo of their organization because of long service, their status as a founder, or because they are a major donor (Arsenault, 1998).

The decision-making process soon after a collaboration may result in behavior such as constant complaining, endless objecting or a kind of generalized anger (Arsenault, 1998). Strongly expressed feelings on one side may evoke equal and opposite reaction on the other. The first source of resistance to a collaboration usually is from board members and comes from the "institutional identification" that a board member feels to his or her old organization. Resistance to the collaboration may also be based on ideology–concern for the "community" or the vague assertion that "their services aren't as good as ours" (McLaughlin, 1998: 230). To this day, board members from the initial two agencies look at the early decision making process differently: on the one hand, one group believes that decisions could have been made quicker and progress would have been faster if there had not been an alliance. Perhaps not surprisingly, the other group believes that the end result was better than if they had all been on the same page initially.

The net effect of this was the sense that, initially, board members did not look at decisions for the health of Food Runners, but voted their allegiance to their "other" organizations. Votes on critical issues were often not taken at all when it appeared that board members would vote the party line (resulting in a 7-7 tie), and one organization accused the other of putting the brakes on anything that would potentially result in a loss of control. Boards of NPOs desperately want not only consensus, but unanimous agreement (McLaughlin, 1998). Consequently, in 2003 the bylaws were changed so that all three organizations could have up to seven board members, bringing the total possible members to 24 (including the three directors who are ex-officio). When a board member resigned, the party that was authorized to appoint that person was responsible for appointing another to fill the vacancy. Twenty-four members is extremely large for a board, whose optimal size is normally nine to 14 members (McLaughlin, 1998), but fairly typical in NPOs, which often head off possible conflicts by creating large boards composed of equal numbers of members from the combining organizations.

After the initial difficulties, a different philosophy about the Food Runners' board has evolved. As new board members are added, they are people dedicated to Food Runners, not to one of the other organizations, and they are being assessed on their skills, not just their fund raising abilities. Whereas each party had initially filled its vacancies with "one of their own," they are now appointing "neutral" parties to the board. Board voting is no longer seen as a problem. Decisions are more neutral now and there is a clearer understanding of what Food Runners is all about. However, governance and decision-making issues that are broader than one organization still need to be worked out.

Culture and Implementation/Integration

Once the decision to jointly build a facility with a kitchen to serve the hunger community had been made, the boards of both organizations met weekly for several months to plan for its construction. The FRC board did not initially see the need for an executive director of Food Runners. A volunteer and a development committee helped with supervising the building construction; however, decisions about the new facility were made by board members who found themselves mired in staff-type functions rather than strategic planning. Opinions about the need for a project manager or FRC executive director differed: MOW wanted to hire one much earlier; IFFS believes that the hiring of the FRC director was done at the time it was needed. The need for a governance structure was also viewed differently. MOW was of the opinion that one should be in place with a more formal business plan; IFFS did not understand how you could create a governance structure if you did not even know what the concept of the building was.

Consequently, because of the need to get the building done, an architect was hired before governance or planning issues were decided. Difficulties with this architect (who was from out of the area), higher-than-expected administrative costs, and process issues caused project overruns. As a result, once the FRC director came on board in November 2003, there was not enough money for operations, and additional funding of $250,000 had to be secured through the bank servicing Food Runners.

As the building neared start of construction, it became obvious to the FRC board that they needed a leader with organizational skills who could get them through the construction of their new building and manage it after completion. In November, 2003, an executive director with previous experience as an engineering manager and with mergers and acquisitions, was hired. The FRC director's charter was to first get the building built and, second, to put the Food Runners organization together and make it operational. The director saw his role as one of facilitating the mission of the other two organizations through collaboration and synergy.

FRC's mission. Initially, Food Runners was formed to be a fund raising entity for the new kitchen facility. Over time, however, the entity developed a mission of its own: landlord for the other two groups and the vendor for senior meal programs. FRC's role was often a point of contention in the beginning. Each agency had different needs they wanted FRC to fill. Despite the frustrations inherent in such a process, the two agencies were determined to stay the course. It was a long learning curve, but it resulted in a stronger collaboration in the end.

Building trust. Trust is one of the most underrated forces in our society, and it is the fabric of nonprofit operations (McLaughlin, 1998); for example, they often get funding without contracts to back it up. It is also key to a successful integration. In a nonprofit collaboration, there may be more enthusiastic sharing of ideas than you would find in a merger; alliances result in new ways of working together without the competition for jobs that often accompanies mergers (McLaughlin, 1998). On the other hand, differences in management style are often reflected in differences in the board, which may cause concern over being able to "trust" one another (McLaughlin, 1998).

Trust is essential to consensus building and to foster effective communication (Giffords & Dina, 2003). It can be compromised by misunderstandings flowing from different cultures or fostered through beliefs that both parties are being honest. Successful alliances respect differences and work through problems as they develop new norms (McCambridge & Weis, 1998). At first, trust between the two agencies caused difficulties. People were unsure how much information to share; board members from one party would question the decisions of the other party; funding sources and methods of operation were also sources of unease. This sense of distrust was dissipated to some extent by the creation of the name of the new organization (Food Runners Collaborative) and by the creation of a logo that incorporated symbols from both organizations. It was ultimately communications, however, that saved the organization.

Communication. To ensure a successful collaboration, information must be shared regularly and in multiple ways (Giffords & Dina, 2003). Adequate time and effort need to go into defining the mission, vision and values of the collaborators. After the decision to build a kitchen facility was made, board members from MOW and IFFS had a retreat to discuss strategic planning. Subsequently, four members from each agency met weekly to plan the building.

Once the FRC director was hired, one of his first initiatives, along with getting the building construction on target, was to have a planning session with the other two executive directors to discuss their three respective roles and the relationship of the three organizations. The initial sentiment was that FRC should fade into the background; however, the final result of the discussions was that they were all three co-equals, three intersecting circles, with no one organization having more "power" than any other. The executive directors and the board went through a "healing" process and communication became more frequent and more open. A decision-making process for "tough" decisions was implemented. Weekly meetings of the three directors where they look for opportunities to work together for overall benefit were also established and continue to this day. All three directors worked hard at establishing communications and allowing for individual styles and personalities.

A Successful Recipe

Meals on Wheels and the Food Shuttle have offices on either side of a building that houses a kitchen facility and the Food Runners' staff between them. Their missions and cultures remain distinct and they even have different pets: MOW has an all-white dog; the Food Shuttle a solid black cat. Despite these differences, however, the collaboration works. Overall, both parties believe the collaboration has turned out to be good for the hunger community, and each organization has changed–for the better–since their alliance. The process was painful, but worth it. As one director said, "[we] made all the mistakes, but it still came out okay." Although there have not yet been as many synergies as they would like, the parties are discussing possibilities such as benefits for the workers. Meals on Wheels has maintained its client base in the face of significant government funding cuts that would have otherwise forced them to remove clients; Food Shuttle has doubled the amount of food that can be gleaned. The FRC board is acting in the best interest of the clients, not the individual agencies; there is not as much "we vs. they" as there used to be; and, Meals on Wheels is trying to expand its sand box.

The true driving force of a collaboration between nonprofits is that the two parties share a similar vision, that the change the organizations are seeking–the problem they are trying to solve–is compatible (La Piana & Hayes, 2005). A key reason in the decision to collaborate is the belief that the alliance will be able to advance the mission of the parties. The reason that this collaboration has worked is the parties' belief in their overall missions of assisting those who are hungry. For nonprofits, the absolute conviction that what they are going through will be good for the community or their customers, or will make the world a better place is what gets them through difficult times (Arsenault, 1998). As one board member responded when asked why they kept at it: "The mission! The mission! The clients! The clients!" Despite all of the difficulties, both organizations kept the vision of alleviating hunger in mind; it was what caused them to come together initially, and it is what keeps them together now.

PRACTICAL LESSONS

Collaborations are successful only when stakeholders expend considerable time and effort to formulate and adopt a culture for the new combined organization (Giffords & Dina, 2003). The culture must fit the context in which the organizations operate in order to be effective (Flannery & Deiglmeier, 1999). In this case study, we examined how two different cultures can learn to work together. Although the integration process is still ongoing, there are many lessons that can be taken away from this collaboration of two disparate nonprofit cultures.

First, to have a successful collaboration, communication is the key. You must work to keep the board and other executive directors informed and involved, and work with the appropriate committees for operational decisions. You must develop language that all parties are comfortable with and that all parties understand. Communication helps build trust; without trust, the collaboration will fail. Clarification of jargon and basic terms would have afforded more efficient communications. By side-stepping this small point the collaboration suffered miscommunication that greatly decreased effectiveness.

Next, combining cultures is not always fatal; an organization can retain its basic values while other aspects of its culture change. The initial parties have both grown and changed and yet have maintained their individual cultures. They have learned from each other. Meals on Wheels has begun exploring alternative meal delivery plans. The Food Shuttle has developed a more formal organizational structure. Both organizations are different, yet their cultures remain remarkably in tact. Additionally, little time was initially spend on learning the collaborator's organizational structures; rather, the focus was on establishing the new organization. Time spent defining the original governance structures would have fostered understanding how each entity functioned and would have smoothed the articulation of the new structure and governance, thus affording the new leadership the ability to implement a comprehensive strategic plan.

Many collaborations fail because of the sense that one party has "taken over." While perhaps not always practical, in this study the development of a third "neutral" entity and the retention of autonomy in the combining organizations went a long way to achieving success. There were no changes in reporting relationships, and the leadership of the two original organizations were cognizant of the reactions that individuals within their entities would have to the collaboration and their roles in helping smooth the transition.

Finally, keep the goal in sight. In this case study, there was no model or business plan to guide the parties. Yet, they both believed that what they were doing was right and they knew where they wanted to end up. It was the getting there that was difficult. To this day, there is disagreement between the parties as to whether a more formal business plan was needed, when the FRC director should have been hired, and whether the governance structure should have been in place before construction of the building began. Yet it didn't (and still doesn't) matter. It worked because they shared the same vision and worked through all their difficulties to achieve that vision. The authors are therefore reluctant to be prescriptive and make suggestions as to the course of action they *should* have followed. The three agencies are where they are now *because* there were no recipe cards to follow.

NOTES

1. Gleaning is the process by which useable food is separated from un-useable food.
2. Safe handling of food is arguably Food Shuttle's most important value. They developed the Good Samaritan law that became the national standard for safe food handling.

REFERENCES

Arsenault, J. 1998. Forging nonprofit alliances: A comprehensive guide to enhancing your mission through joint ventures and partnerships, management service organizations, parent corporations, and mergers. San Francisco: Jossey-Bass.

Bates, G. 1999. Book review: Forging nonprofit alliances. *Journal of Management Consulting, 10*(4): 66.

Boardman, A. E., & Vining, A. R. 2000. Using service-customer matrices in strategic analysis of nonprofits. *Nonprofit Management and Leadership, 10*: 397-420.

Buono, A. F., & Bowditch, J. L. 1989. The human side of mergers and acquisitions: Managing collisions between people, cultures, and organizations. San Francisco: Jossey-Bass.

Cartwright, S., & Cooper, C. L. 1995. Organizational marriage: "Hard" verses "soft" issues? *Personnel Review, 24*: 32-42.

Flannery, D., & Deiglmeier, K. 1999. Leading the social purpose enterprise: An examination of organizational culture. In Social purpose enterprise: Entrepreneurs and venture philanthropy in the new millennium: 1-10. The Robert Enterprise Development Fund.

Giffords, E. D., & Dina, R. P. 2004. Strategic planning in nonprofit organizations: Continuous qualify performance improvement–A case study. *International Journal of Organization Theory and Behavior, 7*(1): 66-80.

Giffords, E. D., & Dina, R. P. 2003. Changing organizational cultures: The challenge in forging successful mergers. *Administration in Social Work, 27*(1): 69-81.

Hiland, M. L. 2003. Nonprofit mergers. *Consulting to Management, 14*(4): 11-14, 60.

La Piana, D., & Hayes, M. 2005. M&A in the nonprofit sector: Managing merger negotiations and integration. *Strategy & Leadership, 33*(2): 11-16.

McCambridge, R., & Weis, M. F. 1998. The rush to merge. *The New England Nonprofit Quarterly*, 6-18.

McClusky, J. E. 2002. Re-thinking nonprofit organization governance: Implications for management and leadership. *International Journal of Public Administration, 25*: 539-559.

McLaughlin, T. A. 1998. Nonprofit mergers and alliances: A strategic planning guide. New York: John Wiley & Sons, Inc.

McLaughlin, T. A. 1996. Seven steps to a successful nonprofit merger. Washington, D. C.: National Center for Nonprofit Boards.

O'Brien, J. E. & Collier, P. J. 1991. Merger problems for human service agencies: A case study. *Administration in Social Work, 15*(3): 19-31.

Olie, R. 1994. International mergers. *Organizational Studies, 15*: 381-405.

Schein, E. H. 1992. Organizational culture and leadership (2d ed.). San Francisco: Jossey-Bass.

Skloot, E. 2000. Evolution or extinction: A strategy for nonprofits in the marketplace. *Nonprofit and Voluntary Sector Quarterly, 29*: 315-324.

doi:10.1300/J369v09n02_07

Server Emotional Experiences and Affective Service Delivery: Mechanisms Linking Climate for Service and Customer Outcomes

Yongmei Liu
Jixia Yang

SUMMARY. A conceptual model is presented that proposes servers' emotional experiences and affective service delivery are mediating mechanisms through which climate for service leads to favorable customer outcomes. Four mechanisms through which organizations with a strong climate for service help their servers to provide a positive affective service delivery are proposed, motivation, capability, carryover, and compensation. Specifically, it is suggested that a strong climate for service gives

Yongmei Liu is a doctoral student, Department of Management, College of Business, Florida State University, Tallahassee, FL 32306-1110 (E-mail: yongmei.liu.0002@fsu.edu).

Jixia Yang is Assistant Professor, Department of Management, City University of Hong Kong, Tat Chee Avenue. Kowloon, Hong Kong (E-mail: mgyang@cityu.edu.hk).

Please direct correspondence to Yongmei Liu at the above address.

The authors wish to thanks Pamela Perrewé and Kevin Mossholder for their excellent suggestions and comments.

Paper submitted to the special issue on organizational behavior and human resource management in foodservice in the *Journal of Foodservice Business Research.*

[Haworth co-indexing entry note]: "Server Emotional Experiences and Affective Service Delivery: Mechanisms Linking Climate for Service and Customer Outcomes." Liu, Yongmei, and Jixia Yang. Co-published simultaneously in *Journal of Foodservice Business Research* (The Haworth Hospitality & Tourism Press, an imprint of The Haworth Press, Inc.) Vol. 9, No. 2/3, 2006, pp. 127-150; and: *Human Resources in the Foodservice Industry: Organizational Behavior Management Approaches* (ed: Dennis Reynolds, and Karthik Namasivayam) The Haworth Hospitality & Tourism Press, an imprint of The Haworth Press, 2006, pp. 127-150. Single or multiple copies of this article are available for a fee from The Haworth Document Delivery Service [1-800-HAWORTH, 9:00 a.m. - 5:00 p.m. (EST). E-mail address: docdelivery@haworthpress.com].

Available online at http://jfbr.haworthpress.com

doi:10.1300/J369v09n02_08

rise to servers' positive emotional experiences at work. Such positive emotional experiences provide the impetus for them to put more effort into affective service delivery (i.e., motivation) and be more capable of doing so (i.e., capability); these positive emotions may also be carried over to the service encounter (i.e., carryover). It is also suggested that in a strong service climate, servers are better able to recharge their emotional energy through organizational and social support (i.e., compensation). Model implications and suggestions for future research are discussed. doi: 10.1300/J369v09n02_08

KEYWORDS. Climate for service, affective service delivery, emotional labor

INTRODUCTION

For most diners, the service received at the table is of prime importance for good service (Barrier, 2004). Therefore, as the immediate interface between a restaurant and the guest, servers play a key role in creating high quality dining experiences for guests. Emotions expressed by servers toward, or in the presence of, customers are an important component of service delivery (Ryan & Ployhart, 2003). More often than not, people favor one restaurant over another not only because the chosen place provides better food quality or a nicer surrounding, but also because they have had good experiences with their servers, and these encounters at the table have made them feel good (Berta, 2004; Mattila, 2001). Such good feelings are directly related to affective service delivery (ASD), namely, the degree to which the service delivery is perceived as friendly and warm (Grandey, 2003).

An accumulation of empirical findings have suggested that servers' positive emotional displays are related to positive customer reactions, including customer satisfaction, loyalty, perception of service quality, and positive word-of-mouth (e.g., Pugh, 2001; Tan, Foo, & Kwek, 2004; Tsai & Huang, 2002). Moreover, as customers become more adept at discerning real and fake emotions, service employees are increasingly expected not only to express positive emotions, but to do so in a sincere manner (Ashforth & Humphrey, 1993; Ferguson Bulan, Erickson, & Wharton, 1997). Together, this attests to the importance of the authentic emotional display that conveys courtesy, friendliness, and warmth during server-guest interactions. Ideal as it is for servers to naturally display the emotions desired, various factors may impede them from doing so. For example, the variety and unpredictability of individual customer demands may make service encounters difficult for servers to manage (Walsh, 2000; Tan et al., 2004), and servers' physical and psychological resources may be depleted when the customer volume is high (cf., Rafaeli & Sutton, 1990). Therefore, servers' emotional states may not always be in tune with the expected emotional expressions. This leads servers to perform emotional labor (Hochschild, 1983), that is, to exert effort so that their emotional expressions and/or experiences will be appropriate for the immediate service situation in order to maintain quality service.

Emotional labor is an effortful process, and prolonged effort in managing one's emotions could have detrimental effects on individuals (Hochschild, 1983; Morris & Feldman, 1996). Indeed, emotional dissonance and emotional exhaustion are common symptoms associated with service work (e.g., Brotheridge & Grandey, 2002; Grandey, 2003). These negative symptoms can further cause servers to have poor service performance and intentions to quit (Cropanzano, Rupp, & Byne, 2003; Cote & Morgan, 2002), which, in turn, can be harmful to service organizations (Kacmar, Andrews, Van Rooy, Steilberg, & Cerrone, in press). Thus, it appears that, to offer ASD to customers consistently in the long run, servers not only need to labor emotionally, they also need to labor well, that is, to perform emotional labor in a healthy way such that the servers' well-being is not jeopardized.

As such, restaurant owners and managers face the challenge of maintaining a high level of customer service and a healthy workforce at the same time. What can restaurant organizations do to balance these two seemingly conflicting goals? Drawing from previous research on climate and emotion, this paper explores the possibility of meeting the challenge through establishing a strong climate for service. Specifically, the paper explores how a climate for service influences servers' emotional experiences at work, which further affects their ASD and positive customer outcomes.

The paper contributes to the literature in at least two ways. First, the "black box" between climate for service and customer outcomes is examined by tapping into server emotions. Climate for service has come to be viewed as a strategic tool for service-based organizations to retain customers and gain profits (Schneider, Bowen, Ehrhart, & Holcombe, 2000). The recognition of the strategic importance of climate for service can be largely attributed to the growing research linking climate for service with higher service quality and greater customer satisfaction (e.g., Johnson, 1996; Pugh, Dietz, Wiley, & Brooks, 2002; Schneider, Ehrhart, Mayer, Saltz, & Niles-Jolly, in press; Schneider, White, & Paul, 1998). This line of research has documented a clear relationship between employee perceptions of a climate for service and customer experiences of service quality (Schneider et al., 2000). What remains less clear, however, is the process through which climate for service leads to positive customer experiences. To date, researchers do not have solid knowledge addressing the question of how climate for service makes a difference in customer perceptions and satisfaction. Considering the role of emotion is critical for understanding this process.

Another contribution of the paper is that it taps into the dynamics of servers' emotional experiences at work, which relate to the authenticity of the servers' emotional display that impacts customer outcomes. Prior research on emotion in the service setting has primarily focused on factors influencing service employees' emotional displays, rather than those affecting their inner emotional experiences. It is noteworthy that, although positive emotional displays in the service encounter are related to positive customer outcomes, the display of positive emotions may come at a cost for the performing employees. In addition, although employers may have the discretion to require their servers to smile to customers, forced smiles are not equally powerful in generating positive customer reactions as authentic smiles (Peccei & Rosenthal, 2000). Thus, the current paper explores the mechanisms through which organizational climate can help servers to minimize costs associated with ASD by influencing the way they experience their emotions and the way they perform emotional labor. Emergent research indicates the importance of warmth and friendliness in service encounters. To extend this research stream, more attention to servers' emotional experiences is needed.

THE PROPOSED MODEL

It is proposed that climate for service provides two primary means through which servers are able to successfully achieve ASD, experiencing positive emotions, and recharging emotional energy (See Figure 1). The ultimate goal is to examine the ways through which such positive influences of climate for service help facilitate employee service performance and favorable customer reactions.

FIGURE 1. The Role of Emotion in the Climate for Service–Customer Outcome Relationship

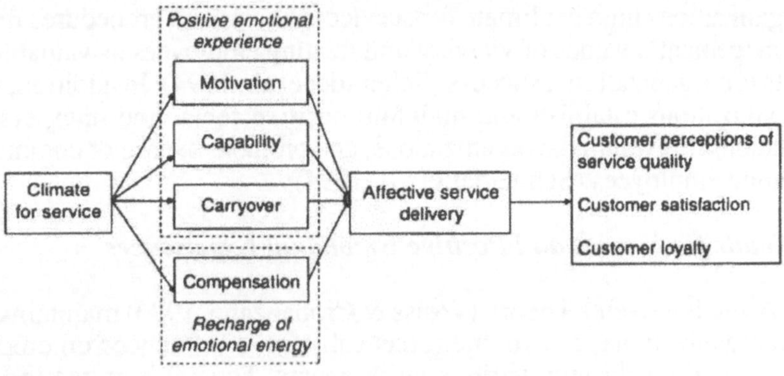

Climate for Service

Climate for service is defined as organization-based employee perceptions of the practices, procedures, and behaviors that are expected, supported, and rewarded with regard to customer service quality (Schneider, Gunnarson, & Niles-Jolly, 1994; Schneider et al., 1998). Two factors are fundamental to the formation of a climate for service. One factor involves general facilitative conditions for service excellence, including efforts toward removing obstacles to work, supervisory behaviors, and human resource policies (Burke, Borucki, & Hurley, 1992; Schneider et al., 1998). This factor reflects a concern for customers. The other factor which reflects a concern for employees is the quality of internal service among units within an organization. For example, an organization with a strong climate for service stresses the importance of service throughout the entire organization, not just for service employees in physical contact with customers (Schneider & Bowen, 1993). Such an organization is characterized with smooth inter-functional coordination, open communication among units, interpersonal warmth and friendliness, and pro-social behaviors targeting at the common good of the organization (Ostroff, 1993; Schneider et al., 1994).

A central characteristic of climate for service is an organization's strong concern for its employees (Ryan & Ployhart, 2003; Schneider & Bowen, 1993; Schneider et al., 1994). In organizations with strong service climate, employees are well-trained, and attention is given to make sure that employees are supported with up-to-date and well-serviced equipment and supplies. Schneider and Bowne (1993) stated that a climate for employees' well-being is the foundation for service climate. In organizations high in climate for service, policies and procedures reflect management's values of viewing and treating employees as valuable assets for organizational success (Schenider et al., 1994). In addition, such organizations establish and maintain positive social and interpersonal relationships within the organizations, and promote a sense of community among employees (Schneider et al., 1994).

Climate for Service and Positive Emotional Experiences

Affective Events Theory (Weiss & Cropanzano, 1996) maintains that the organizational environment has substantial influences on employees' emotions through various work events. Thus, it is suggested that the prevalence of a climate for service constitutes an important facet of the work environment that greatly shapes service employees' emotional experiences. Particularly, it is argued that, compared to restaurants with a weak climate, restaurants with a strong climate for service evoke more recurring affective events which induce positive emotional experiences among servers.

A positive emotional climate has been associated with an organization's attentions to the emotional needs of employees (Ozcelik, Langton, & Aldrich, 2001). When employees view themselves as being treated well by the organization and enjoy their relationships with their supervisors and peers, they tend to evaluate their work in a positive way (Schneider et al., 1994), which in turn leads to pleasant feelings, such as joy, pride, and contentment (Lazarus, 1991; Watson, 2000). For example, service-oriented organizations generally emphasize employee training, performance appraisals, and pay equity (Schneider, Wheeler & Cox, 1992). As a result, employees may feel pleased when presented with opportunities to participate in training programs that enhance performance knowledge and skills, when shown the importance of customer service, and when given appreciation for excellent service work.

Working in an organization that is strong in climate for service, individuals are likely to build confidence, self-worth and respect through a variety of experiences, which in turn give rise to positive feelings among employees. Consistently, it has been found that a climate for service is positively related to service employees' sense of self-determination and professional competencies through the influence of empowerment leadership behaviors, such as participation in decision-making and information sharing (Yagil & Gal, 2002). Such positive self-evaluations have also been found to lead to positive emotions (Isen, 1999).

Further, the work relationship is an important aspect of an employee's emotional life at work. A sense of belonging and not alone in one's social networks at work is associated with a sense of strong attachment to the organization (Kahn, 1998). Because positive, nourishing relationships among employees are encouraged and cultivated in organizations high in climate for service (Ostroff, 1993), employees in such organizations enjoy a relatively strong sense of attachment (Carr, Schmidt, Ford, & DeShon, 2003), which tends to induce further positive emotions (Lawler, 1992). Consistently, it has been found that the more service employees seem to enjoy interacting with other employees at work, the more likely they have positive feelings about work (Ferguson Bulan et al., 1997).

Finally, within a climate for service, employees tend to identify with its strong orientation toward customer service (Schneider et al., 1998). Driven by this internalized value, employees strive for a high quality of customer service (Peccei & Rosenthal, 2000). The process of embracing and internalizing values and goals pertaining to customer service is enmeshed with emotions. Individuals seek to experience certain fundamental sentiments through their behaviors to confirm their social identity (Ashforth & Saks, 2002). For service employees who resonate with the role of serving others, they not only think but experience emotions in the identity construction process. Especially when the work environment is characterized by a strong customer orientation, employees tend to more diligently integrate work role requirements with authentic emotions.

Through the emotional socialization of using affective reactions to gauge one's emotional resonance with the work role, even short-term inauthenticity may lead to longer-term authenticity (Ashforth & Saks, 2002). As a result, service employees having salient personal identities with their service roles may tend to experience authentic positive feelings that are consistent with display rules (Ashforth & Humphrey, 1993). As indirect support to this argument, commitment to display rules has been found to facilitate deep acting and ASD (Gosserand & Diefendorff, 2005), suggesting that identification with the service role not only motivates the service employees to try to experience emotions that they express, but also is successful at doing so. Elsewhere, it was found that perceived demands for positive emotional expressions were associated with physical symptoms for those who had low organizational identification, but not for those who highly identified with their organizations

(Schaubroeck & Jones, 2000). This suggests that organizational identification promotes positive emotions, which makes expressing positive emotions less stressful for service employees.

In addition, restaurants with a strong climate for service are often well-known and applauded by the public. The organization and associated brand name(s) are well received among patrons. Such a corporate reputation may also cause employees to feel positive emotions, such as pride and inspiration (cf. Dutton & Dukerich, 1991). As indirect support to this argument, Ryan, Schmit and Johnson (1996) found that overall customer satisfaction positively influenced the morale of service employees.

In sum, it is argued that servers tend to experience more positive emotions at work in organizations high in climate for service. The positive emotions may stem from work events, work relationships, internalization of organizational goals and values, and organizational membership. Below it is further argued that servers' positive emotional experiences at work will influence their ASD at service encounters through three mechanisms labeled motivation, capability and carryover.

Motivation. Work motivation determines one's choices of whether to engage in a certain activity, how much effort to exert and how long to persist in that activity. Gosserand and Diefendorff (2005) found that individuals must be committed to emotional display rules for such rules to impact actual behavior, suggesting that motivation plays a role in ASD. In this paper, motivation is defined as the process in which service employees willingly put forth the effort needed to maintain a high quality service standard. It is argued that positive emotional experiences serve as an impetus motivating servers to better manage service encounters.

Emotions may influence motivation through expectancies (Erez & Isen, 2002; George & Brief, 1996; Seo, Feldman Barrett, & Bartunek, 2004). People in a positive emotional state have a more positive outlook for future events, and tend to focus more on potential positive outcomes (Fredrickson, 1998; George & Brief, 1996; Staw & Barsade, 1993). For this reason, they also tend to set difficult goals, be persistent with the goals, and exert more effort toward achieving the goals (Seo et al., 2004). In their experimental studies, Erez and Isen (2002) found that individuals who were in a positive emotional state were more likely to believe that their efforts would result in a desired performance, and they also had a higher level of enjoyment of the favorable rewards obtained. Accordingly, it is proposed that servers who feel positive emotions are more likely than others to feel that their efforts at serving customers will achieve customer satisfaction and are more likely to be proud that they are able to achieve the goal of quality customer service.

In addition, positive emotional states generally encourage cooperation and reduce aggression in interpersonal interactions (Isen & Baron, 1991). Largely due to the motivation of mood maintenance, individuals experiencing positive emotions are more empathetic and willing to help others (George & Brief, 1992; Staw, Sutton, & Pelled, 1994). They also tend to be more cooperative when conflicts arise (Isen & Baron, 1991). As such, being in a positive emotional state, servers will be less likely to be annoyed by minor problems in customer encounters. Rather, they will tend to be more entertaining, understanding and forgiving. Such a stable, positive emotional state will help them establish personal connections with customers, which enhances customer loyalty. Servers' stable, positive emotions may even generate a positive spiral in the server-patron emotional exchange process (Staw et al., 1994), which makes the service encounter more genuinely enjoyable for patrons and servers, further motivating servers to deliver better service in the future.

Thus, service employees who experience positive emotions are motivated to anticipate customer needs and are ready to help them fulfill those needs. By doing so, positive emotional expressions are fostered more easily in server-patron interactions. This will ultimately give rise to successful ASD, which enhances customer satisfaction and loyalty.

Capability. Capability is defined in this paper as the knowledge, skills and abilities of servers to carry out high standard service. Servers need not only to be motivated, but also to have the capability to serve guests in an appropriate way. Thus, other things being equal, servers in a positive emotional state should be more capable of managing various situations and more flexible in solving customer problems.

Emerging research on positive psychology has linked positive emotions with behaviors that are approaching, exploring, learning and creating in orientation (Fredrickson, 1998, 2001). Whereas negative emotions function to narrow a person's momentary thought-action repertoire, experiences of positive emotions prompt individuals to discard routine-type behavioral scripts and pursue novel paths of thought and action. It also has been found that individuals in positive emotional states are more flexible and creative in thinking and problem solving (Isen, 1999). In the service setting, servers in positive emotional states may see various aspects of a situation, be ready to understand customer needs and flexible in finding ways to satisfy those needs. This also means that when problems arise, they will be proactive and skillful in managing them so as to maintain a friendly atmosphere.

Moreover, positive emotions help individuals quickly recover from repercussions of a negative event (Fredrickson & Levenson, 1998), and thus should aid service employees in recovering from negative mood states caused by daily work hassles or negative encounters with difficult customers. Service employees could then focus on serving customers in a manner prescribed by the organizational rules and policies, thereby helping to achieve ASD. Thus, positive emotions help servers to perform at their highest potential, which further gives rise to better ASD.

Whereas extensive research has documented the effects of positive emotional states on performance through motivation and increased capability (e.g., Erez & Isen, 2002; George & Brief, 1996), little research evidence is available in the service context. However, some indirect evidence does exist. For example, positive affectivity has been found to be positively related to deep acting and negatively to surface acting (Gosserand & Diefendorff, 2005). Assuming service providers with positive affect are likely to express positive emotions frequently in customer encounters (Schauboeck & Jones, 2000); this suggests that the positive emotions of service providers enhance their motivation to perform affectively. Elsewhere, it has been found that positive customers (as indicated by positive personality traits or emotional states) were associated with performance gains of service providers (Van Dolen, de Ruyter, & Lemmink, 2004; Tan et al., 2004), which suggests that positive interpersonal encounters that give rise to positive feelings of servers enhance their motivation and capability in ASD. Such benefits resulting from positive interpersonal dynamics should also accrue when it occurs among employees. For instance, it has been found that a service emphasis had a positive impact on employees' perceptions of service capability, which was further associated with employees' perceived service quality, indicating their positive evaluation of their own service performance and satisfaction of the customers (Gupta, 1998). Such influence of a service emphasis on employees' perceived performance is likely to occur because of the positive emotions induced by the climate for service, which enhances the service providers' motivation and capability to perform well in customer encounters.

Carryover. Carryover effects occur when emotions generated in one domain and by events of that domain carry over to be experienced in another domain (Edwards & Rothbard, 2000). For example, it has been suggested that emotions at work and emotions at home are likely to be entangled and influence each other (e.g., Edwards & Rothbard, 2000;

Wharton & Erickson, 1993), indicating the carryover effects between the work and family domains. Given the focus of this paper, carryover is considered the process in which positive emotions induced by the internal work environment are carried over to service encounters. It is suggested that, given the proximity between the internal work environment and the service encounter, service employees are subject to the carryover effects when they juggle between the role of an employee and the role of a service provider.

Some may argue that service employees may treat work relationships and service encounters in psychologically separate accounts, and therefore, positive emotions generated from the internal work environment may not significantly influence their emotions in service encounters, that is, the carryover effects may not occur. However, this paper argues that the opposite case is true for two reasons. First, service employees hold boundary spanning roles (Ryan & Ployhart, 2003), and boundaries between the two domains may not always be clear-cut for service employees. Due to the close contact with various customers day by day, service employees develop a deep understanding of customers' needs and perspectives and become synthesized with them. Thus, service employees are likely to view both coworkers and customers as integral parts of their work, and do not alter their general emotional behavioral patterns when interacting with them. Second, being in positive emotional states may make servers view the internal and external work environment as more of a whole, rather than consciously distinguishing coworkers and customers as two separate groups of people. In this regard, it has been suggested that positive emotions tend to draw people closer (Wharton & Erickson, 1993), and it has been found that individuals in positive emotional states are more likely to classify peers of an out-group as in-group members (Dovidio, Gaetner, Isen, & Lowrance, 1995).

Rothbard's (2001) engagement hypothesis postulates that positive emotions experienced at work can spill over to the home domain. Following this logic, it is posited that positive emotions experienced by service employees in the interactions with their co-workers at the "back-stage" can be carried over into the "front stage" where servers interact with patrons. As indirect support to this argument, Rothbard and Wilk (2004) found that call center workers' mood at the beginning of the day, which is most likely determined by the inner work domain, such as one's work relationships, influenced the emotions they expressed when handling customer calls.

Climate for Service and the Recharge of Emotional Energy

Despite the positive influence of climate for service on employees' emotional experience at work, service employees may still feel, from time to time, that they are emotionally drained by their job demands (Brotheridge & Lee, 2002). In such situations, it is critical that they have opportunities to recharge their emotional energy. Emotional energy refers to the level of confidence, warmth, and enthusiasm one feels in interpersonal encounters (Collins, 1981). Individuals who possess an adequate level of positive emotional energy have greater emotional reserves to draw upon when interacting with others and dealing with interpersonal problems that arise. Emotional energy is a valuable resource for employees who are constantly interacting with customers. In service encounters, however, servers may lose their emotional energy due to various reasons such as high work loads, demands from difficult customers and so forth (e.g., Pugh, 2001). Individuals who lose emotional energy tend to experience negative interpersonal outcomes such as rejection (Collins, 1981). The section below discusses how organizations high in climate for service, provides their employees with opportunities to recharge their emotional energy. In this paper, this provision to employees is termed *compensation.*

According to the Conservation of Resources Theory, a loss of valued resources is stressful in and of itself, especially when it initiates a loss cycle (Hobfoll & Shirom, 2000). Service performance will not measure up to organizational expectations when employees' emotional resources are constantly tapped without being replenished.

One effective way for workers to regain emotional reserves is through positive, rewarding social interactions with their constituents such as clients and coworkers (Brotheridge & Lee, 2002). Although servers may be emotionally uplifted by positive interactions with guests (cf., Van Dolen et al., 2004; Tan et al., 2004), it is likely that servers primarily rely on co-workers for regaining their emotional energy. The reason is that the recharge of emotional energy via interpersonal processes requires mutual emotional sharing, and oftentimes, the release of tension by venting emotional experiences inconsistent with emotional norms (Thoits, 1996). To accomplish this, one has to enjoy an adequate degree of openness and freedom in the emotional exchange. Although some experienced servers are able to build long-term personal relationships with their patrons, most server-patron interactions remain short in duration and relatively nonpersonal (Kruml & Geddes, 2000), which does not allow an adequate level of closeness to foster emotional openness.

In addition, service employees face many restrictions in what they can communicate to customers due to implicit power differences between the two as well as explicit organizational policies and rules. Therefore, to compensate the emotional energy lost during service encounters, employees largely depend on the rewarding social relations with constituents from within the organization.

Consistent with the above-mentioned view, it has been found that whereas the amount of time spent with customers is related to service employees' feelings of inauthenticity and fewer positive emotions about their work; the amount of time spent with coworkers at work tends to have positive effects on employees' well-being (Ferguson Bulan et al., 1997). This is partly because the change of scenario allows service employees a mental break from the service role and an opportunity to enjoy and appreciate the positive aspects of their work, both of which have been found to help replenish one's affective resource (cf., Frankenhaeuser, Lundberg, Fredrikson et al., 1989; Fritz & Sonnentag, 2004–as cited in Beal, Weiss, Barros, & MacDermid, 2005).

A climate for service entails organizational policies and practices that reward and support superior customer service. The general facilitative conditions enabled by climate for service prevent employees' energy, resources and time from being depleted by internal politics, so that they can dispense their emotional energy in service encounters. Further, a climate for service is associated with positive interpersonal relationships among employees (Schneider et al., 1994). Such rewarding social interactions with coworkers serve as the most reliable source that service employees can rely on to recharge their emotional energy depleted in service delivery. Consistently, it has been found that more social interactions at work positively influence one's emotional states and general job satisfaction (Tschan, Semmer, & Inversin, 2004; Ilies, Johnson, & Judge, 2005). Instrumental and emotional support from one's peers and one's supervisor has been found to make employees work harder and have a positive orientation toward problem solving, which further buffer the impact of work overload and decreases the experiences of emotional exhaustion (Bakker, Demerouti, & Euwema, 2005; Ito & Brotheridge, 2003). Moreover, assistance and help from coworkers contribute significantly to service providers' customer orientation and subsequent customer perceptions of service quality (Schneider et al., 1998; Susskind, Kacmar, & Borchgrevink, 2003). In summary, climate for service affects employee ASD positively by recharging servers' emotional energy to be utilized in service encounters.

DISCUSSION AND CONCLUSION

Loyal patrons are much more valuable to a restaurant than just casual diners (Mattila, 2001). Like other service-based organization, to attain and maintain customer satisfaction and loyalty, restaurants have been promoting "service with a smile" for years (Mattila, 2001). However, less attention has been given to the understanding that it is a genuine smile not a fake smile that matters to customers. Indeed, there is room for restaurants to improve in order to facilitate servers' genuine smiles. This paper seeks to identify one potential area for restaurants to promote genuine smiles to customers so as to keep customer satisfaction and loyalty. The process through which climate for service is transformed to favorable customer responses is also explored.

It is argued that emotion serves as a transformational basis upon which climate for service is linked with employee behavior and customer outcomes. The conceptual model suggests that climate for service makes up an organizational context conducive to servers' ASD, which further leads to desirable customer outcomes such as satisfaction and loyalty. Specifically, four mechanisms–motivation, capability, carryover, and compensation–are discussed, each of which help explain how an organization high in climate for service is able to help its service employees to enjoy an emotional life at work that is of a high quality. Motivation and capability concern how service employees who enjoy positive emotional experiences, induced by a strong climate for service, tend to perform better in customer interactions. Carryover refers to situations when the service employees' positive emotional experiences induced by their internal work environment carry over to customer interactions. Compensation occurs when one's emotional energy is recharged by reinforcing social interactions and organizational policies that cultivate concerns for customers and employees alike.

In short, it is suggested that in organizations high in climate for service, service employees tend to stay in more positive emotional states in general, which lead to their more frequent and more genuine positive emotional expressions in service encounters. When service employees do experience emotional drain, a climate for service allows them to be able to quickly and adequately recharge their emotional energy. Past research has suggested that emotional labor could be detrimental to service employees' well-being (e.g., Hochschild, 1983; Kruml & Geddes, 2000), and that failing to labor well with emotions (e.g., being perceived as inauthentic by customers) may negatively influence customer perceptions of service quality (Ashkanasy, Härtel, & Daus, 2002; Rafaeli

& Sutton, 1987). The model proposed in this paper suggests that, by building a strong organizational climate for service, an organization may be able to help its service employees go through their difficulties in the process of ASD, which further brings forth positive customer reactions.

Model Implications

The implications of the model are many. First, the model has important implications for research on service climate. Due to the dominance of the cognitive perspective in the service climate research, the literature discusses little concerning the emotional base that underlies organizational climate for service. This paper works toward this direction by demonstrating that climate for service can not only be perceived, but also be felt, by employees; and that the feelings induced by the climate may actually make a difference in their service work.

Moreover, although research on service climate has been advanced by knowing more about what climate for service means for customers, it is equally important to know what climate for service means to employees (Peccei & Rosenthal, 2000), especially in an emotional sense, given the affective nature of the service work. This paper also points to the need for understanding what emotional impact service climate might particularly have on service employees' ASD. The model broaches this research subject by examining how climate for service might influence employees' emotional experiences at work, which in turn enhances employees' ASD. It is suggested that emotion serves as one of the linking pins that draw service climate, employee behavior, and customer reactions together. Although past research in the service setting has drawn a clear link between climate for service and customer reactions, the process through which such a link is present remains unclear. Consistent with the recent trend in the emotion literature, it is argued that cognition alone cannot provide complete explanations for organizational phenomena. Thus, to move the linkage research relating service climate to customer outcomes forward, the role of emotion in the process should be fully considered. By focusing on emotional processes, however, it is not the intent to argue that the cognitive and behavioral processes are less important. It is most likely that these processes intertwine and mutually influence each other. Future research needs to incorporate both affective and non-affective processes in developing a comprehensive model of the process through which service for climate influences customer outcomes via employee values, attitudes, emotions and behaviors.

The model also has implications for research on emotional labor. In the emotional labor literature, prior research has mainly focused on the influences of individual difference variables on one's choice of emotional labor strategies (e.g., Brotheridge & Grandey, 2002; Schaubroeck & Jones, 2000). Nevertheless, researchers do not yet understand how the organizational context influences individuals' emotional labor performance. Recent studies have started to consider some physical environmental factors such as store busyness (Pugh, 2001; Tan et al., 2004). However, contextual variables such as organizational climate may warrant particular research attention. The impact of more "soft" organizational contexts, such as climate for service, on emotional labor is supposed to be more subtle and complex in that such contexts in and of themselves are very psychological and emotional. It is the hope that this paper will stimulate more research looking into the contextual influences on service employees' emotional labor. For example, what emotional labor strategies do service employees opt to use when a certain contextual attribute is salient in the work environment? Or what are the differential effects on emotional labor strategies if multiple contexts are prominent at the same time in the workplace? Clearly, more research is needed to have a better understanding of these questions.

Lastly, the model has implications for human resource management practice. Guided by the widely held belief that customers react positively to smiling and friendly service employees, organizations have rushed to develop human resource practices that either attempt to select employees with the "right" type of personality, or to train and socialize them to adhere to the emotional display rules prescribed by the organization (Callaghan & Thompson, 2002). However, regardless of what personality traits one may possess or what emotional labor strategies one may adopt (e.g., surface acting or deep acting), performing emotional labor still can deteriorate one's health (Schaubroeck & Jones, 2000). This paper suggests an alternative route that an organization striving for excellent service could take, that is, to develop and strengthen a climate for service. By building a climate for service in which employees' well-being is enhanced, employees' contributions are recognized and appreciated, and supportive working relationships are available, organizations may be better able to help their service employees manage the emotional demands in service encounters.

Therefore, rather than simply investing in selection, training and socialization in the hope of more smiles on the faces of service employees, organizations may also develop human resource practices that evoke hope, fun, excitement and resilience in the hearts of the employees, which provides them the needed emotional energy to deliver affective service effectively. This is not, however, to disregard the recommendation suggested by many authors (e.g., Arvey, Renz, & Watson, 1998) that having the right people with the needed skills to deliver the service is important. Intensive selection procedures do help prevent misfit between employees' emotional disposition and the affective demands at work, and training and socialization do hold potential for enhancing one's capacity for handling difficult situations. However, what has to be emphasized is that service employees have to internalize the values conveyed by the emotional norms in service organizations in order for them to labor well with their emotions. Thus, selection, training and socialization need to go hand in hand with fostering the right climate. An organization focusing on one factor while overlooking the other runs the danger of underperformance from its service employees.

Future Research

Looking to the future, many research questions are open for inquiry. First, the model needs to be tested. It will not be an easy task, as the nature of the model requires both experience sampling techniques and matched survey design. Specifically, to assess the occurrence of carry-over and compensation, employees' emotional experiences and social interaction activities at work will have to be sampled multiple times daily over a certain period of time. To minimize the bias associated with common method variance, it is suggested that managers or peers rate the focal employees' motivation, capability and levels of ASD and customers provide assessment of service performance and their own attitudinal reactions to service transactions. Given that servers at restaurants usually work at a fast and irregular pace during their shifts, and that the research design would require customer involvement, it might be difficult to get restaurant organizations committed to such a data collection effort. However, the results will provide useful insights and guidance for both future research and business practices.

Second, the role of emotional intensity of service employees' emotional life deserves more attention in future research. Besides the hedonic quality (i.e., positive vs. negative) of emotion, the intensity of emotional experiences may also influence the degree to which the above mentioned motivation and capability effects occur. An important function of emotion is to signal situations of personal relevance and thus pull the needed attention and create a state of action readiness (Frijda, 1988). However, stimuli that engender only a low level of emotional arousal are not likely to enter into the cognitive appraisal process (Lazarus, 1991), failing to generate any action tendencies. Thus, as compared to emotions of low intensity, emotions of moderate and high intensity are more likely to draw individuals' attention and influence their courses of action. Accordingly, motivation and capability effects should be more salient when emotional intensity is relatively high.

As a side note, to what degree an organization wants its service employees to be emotionally aroused at work remains as an empirical question. Kaufman (1999) proposed an invert-U shaped relationship between emotional intensity and performance. Specifically, he argued that emotional intensity is facilitative to human performance because it helps individuals develop effective coping strategies. However, such an effect occurs only to a certain point. After that point, increases in emotional intensity cause deterioration in human performance. Considering this in the service contexts, customers who find a service worker sad or sleepy may have lower perceptions of service quality, satisfaction, and loyalty; however, one that is too joyful and enthusiastic may lead to negative outcomes. Thus, research needs to be conducted to help organizations build the right type of service climate with a certain level of emotional arousal that is both functional in terms of objective performance and social appropriateness. This is possible through the collective emotion regulation process (George, 2002).

Third, the model has focused on how climate for service enhances ASD through fostering employee positive emotional experiences at work. Yet, the role of negative emotions also warrants attention. To pursue a high quality of customer service, management may reward employee efforts toward this goal as well as punish employee behaviors that hinder the goal attainment. Thus, service employees may have both positive and negative emotional experiences resultant from serving performance. On one hand, they feel encouraged to live up to the performance of ASD. On the other hand, being pressed for excellent customer service in these organizations, employees may feel shameful, sad or guilty for having ruined a particular service transaction. Some research questions regarding this matter include: How should negative emotions of servers be handled? Can negative emotions (e.g., fear and shame) in fact prompt service employees to have good service delivery? Answers to such questions may help service organizations in their efforts to promote a healthy climate for service.

Fourth, this model spelled out the influence of organizational context (i.e., climate for service) on emotional labor. It may be fruitful for future studies to consider both individual differences and organizational context antecedents in search for the best conditions for ASD. Furthermore, it may well be that the positive influences of climate for service on employees are bound by individual differences. For instance, individuals high on self-monitoring may make them be more receptive to the influence of climate for service, as compared to their low self-monitoring counterparts. Investigating individual differences as a boundary condition may prove to be helpful for a richer understanding of the climate-ASD relationship.

Finally, another rewarding line of research is to examine the climate for service at the organizational level. There has been scant research taking a more macro approach to understanding emotion in organizational life. Currently, little attention has been devoted to emotional phenomena at the organizational level, particularly when the context is characterized by climate for service. Some research questions related to this matter include: Are there any basic emotional tones across all the service organizations high in climate for service? If so, how do the emotions come to be shared in the workplace? How do we measure the emotional undercurrents that exist in a climate for service? How does a climate for service differ from a climate that is defined as purely emotional? Indeed, there seems to be a whole new territory for research on emotion and service climate at the organizational level.

In conclusion, this paper presents a conceptual model that connects climate for service with favorable customer reactions through the mediation of ASD. The climate-service delivery linkage and the service delivery-customer outcome linkage were explained by examining the role of emotion. Specifically, four mechanisms through which emotion functions to convert climate for service to employees' actual behavior (i.e., ASD) were discussed. Research on climate for service has been around for years. However, when emotion is integrated into this research area there are many intriguing questions open for future exploration. It is the hope that the current paper made a step forward in stimulating more scholarly interest in this new line of inquiry.

REFERENCES

Arvey, R. D., Renz, G. L., & Watson, T. W. (1998). Emotionality and job performance: Implications for personnel selection. *Research in Personnel and Human Resources Management, 16*, 103-147.

Ashforth, B. E., & Humphrey, R. H. (1993). Emotional labor in service roles: The influence of identity. *Academy of Management Review, 18*, 88-115.

Ashforth, B. E., & Saks, A. M. (2002). Feeling your way: Emotions and organizational entry. In R. G. Lord, R. J. Klimoski, & R. Kanfer (Eds.). *Emotions in the workplace: Understanding the structure and role of emotions in organizational behavior* (pp. 331-369). San Francisco: Jossey-Bass.

Ashkanasy, N. M., Härtel, C. E. J., & Daus, C. S. (2002). Diversity and emotion: The new frontiers in organizational behavior. *Journal of Management, 28*, 307-338.

Bakker, A. B., Demerouti, E., & Euwema, M. C. (2005). Job resources buffer the impact of job demands on burnout. *Journal of Occupational Health Psychology, 10*, 170-180.

Barrier, B. (2004). Study: Operators, guests differ on important service drivers. *Nation's Restaurant News, 38*, 18.

Beal, D. J., Weiss, H. M., Barros, E., & MacDermid, S. M. (2005). An episodic process model of affective influences on performance. *Journal of Applied Psychology, 90*, 1054-1068.

Berta, D. (2004). Table talk: Smart servers avoid taboo topics with guests. *Nation's Restaurant News, 38*, 50.

Brotheridge, C. M., & Grandey, A. A. (2002). Emotional labor and burnout: Comparing two perspectives of "people work. *Journal of Vocational Behavior, 60*, 17-39.

Brotheridge, C. M., & Lee, R. T. (2002). Testing a conservation of resources model of the dynamics of emotional labor. *Journal of Occupational Health Psychology, 7*, 57-67.

Burke, M., Borucki, C., & Hurley, A. (1992). Reconceptualizing psychological climate in a retail service environment: A multiple stakeholder perspective. *Journal of Applied Psychology, 77*, 717-729.

Callaghan, G., & Thompson, P. (2002). "We recruit attitude": The selection and shaping of routine call center labor. *Journal of Management Studies, 39*, 233-254.

Carr, J. Z., Schmidt, A. M., Ford, J. K., & DeShon, R. P. (2003). Climate perceptions matter: A meta-analytic path analysis relating molar climate, cognitive and affective states, and individual level work outcomes. *Journal of Applied Psychology, 88*, 605-619.

Collins, R. (1981). On the microfoundations of macrosociology. *American Journal of Sociology, 86*, 984-1014.

Cote, S., & Morgan, L. M. (2002). A longitudinal analysis of the association between emotion regulation, job satisfaction, and intentions to quit. *Journal of Organizational Behavior, 23*, 947-962.

Cropanzano, R., Rupp, D. E., & Byrne, Z. S. (2003). The relationship of emotional exhaustion to work attitudes, job performance, and organizational citizenship behaviors. *Journal of Applied Psychology, 88*, 160-169.

Dovidio, J. F., Gaertner, S. L., Isen, A. M., & Lowrance, R. (1995). Group representations and intergroup bias: Positive affect, similarity, and group size. *Personality and Social Psychology Bulletin, 21*, 856-865.

Dutton, J. E., & Dukerich, J. M. (1991). Keeping an eye on the mirror: Image and identification in organization. *Academy of Management Journal, 34*, 517-554.

Edwards, J. R., & Rothbard, N. P. (2000). Mechanisms linking work and family: Clarifying the relationship between work and family constructs. *Academy of Management Review, 25*, 178-199.

Erez, A., & Isen, A. M. (2002). The influence of positive affect on the components of expectancy motivation. *Journal of Applied Psychology, 87*, 1055-1067.

Ferguson Bulan, H., Erickson. R. J., & Wharton, A. S. (1997). Doing for others on the job: The affective requirements of service work, gender, and emotional well-being. *Social Problems, 44*, 235-256.

Frankenhaeuser, M., Lundberg, U., Fredrikson, M., Melin, B., Tuomisto, M., Myrstern, A., Hedman, M., Bergman-Losman, B., & Wallin, L. (1989). Stress on and off the job as related to sex and occupational status in white-collar workers. *Journal of Organizational Behavior, 10*, 321-346.

Fredrickson, B. L. (1998). What good are positive emotions? *Review of General Psychology, 2*, 300-319.

Fredrickson, B. L. (2001). The role of positive emotions in positive psychology: The broaden-and-build theory of positive emotions. *American Psychologist, 56*, 218-226.

Fredrickson, B. L., & Levenson, R. W. (1998). Positive emotions speed recovery from the cardiovascular sequence of negative emotions. *Cognition and Emotion, 12*, 191-220.

Frijda, N. H. (1988). The laws of emotion. *American Psychologist, 43*, 349-358.

Fritz, C., & Sonnentag, S. (2004). *Recovery, health, and job performance: Effects of weekend experiences*. Unpublished manuscript, Technical University of Braunschweig, Germany.

George, J. M. (2002). Affect regulation in groups and teams. In R. G. Lord, R. J. Klimoski. R. Kanfer (Eds.), *Emotions in the workplace: Understanding the structure and role of emotions in organizational behavior* (pp. 182-217). San Francisco, CA: Jossey-Bass.

George, J. M., & Brief, A. P. (1992). Feeling good-doing good: A conceptual analysis of the mood at work-organizational spontaneity relationship. *Psychological Bulletin, 112*, 310-329.

George, J. M., & Brief, A. P. (1996). Motivational agendas in the workplace: The effects of feelings on focus of attention and work motivation. *Research in Organizational Behavior, 18*, 75-110.

Gosserand, R. H., & Diefendorff, J. M. (2005). Emotional Display Rules and Emotional Labor: The Moderating Role of Commitment. *Journal of Applied Psychology, 90*, 1256-1264.

Grandey, A. A. (2003). When "the show must go on": Surface and deep acting as determinants of emotional exhaustion and peer-rated service delivery. *Academy of Management Journal, 46*, 86-96.

Gupta, A. (1998). *The relationship between employee perceived service climate and customer satisfaction*, Dissertation Abstracts International Section A: Humanities and Social Sciences, Vol. 59(6-A), pp. 2098.

Hobfoll, S. E., & Shirom, A. (2000). Conservation of resources theory: Applications to stress and management in the workplace. In R. T. Golembiewski (Eds.), *Handbook of organizational behavior* (pp. 57-80). New York: Marcel Dekker.

Hochschild, A. R. (1983). *The managed heart: Commercialization of human feeling*. Berkeley: University of California Press.

Ilies, R., Johnson, M. D., & Judge, T. A. (2005). *Social interactions at work: Their influence on affective experiences and job satisfaction.* Paper presented at the Society of Industrial and Organizational Psychology annual conference, Los Angeles.

Isen, A. M. (1999). On the relationship between affect and creative problem solving. In S. W. Russ (Ed.), *Affect, creative experience, and psychological adjustment* (pp. 3-17). Philadelphia, PA: Taylor & Francis.

Isen, A. M., & Baron, R. A. (1991). Positive affect as a factor in organizational behavior. In B. M. Staw & L. L. Cummings (Eds.), *Research in Organizational Behavior* (Vol. 13, pp. 1-53). Greenwich, CT: JAI Press.

Ito, J. K., & Brotheridge, C. (2003). Resources, coping strategies, and emotional exhaustion: A conservation of resources perspective. *Journal of Vocational Behavior, 63*, 490-509.

Johnson, J. W. (1996). Linking employee perceptions of service climate to customer satisfaction. *Personnel Psychology, 49*, 831-851.

Kacmar, K. M., Andrews, M. C., van Rooy, D., Steilberg, R. C., & Cerrone, S. (in press). Sure everyone can be replaced...but at what cost? Turnover as a predictor of unit-level performance. *Academy of Management Journal.*

Kahn, W. A. (1998). Relational systems at work. In B. M. Staw & L. L. Cummings (Eds.), *Research in organizational behavior* (Vol. 20, pp. 39-76). Greenwich, CT: JAI Press.

Kaufman, B. E. (1999). Emotional arousal as a source of bounded rationality. *Journal of Economic Behavior and Organization, 38*, 135-144.

Kruml, S. M., & Geddes, D. (2000). Exploring the dimensions of emotional labor: The heart of Hochschild work. *Management Communication Quarterly, 14*, 8-49.

Lawler, E. J. (1992). Affective attachments to nested groups: A choice-process theory. *American Sociological Review, 57*, 327-339.

Lazarus, R. S. (1991). *Emotion and adaptation.* New York: Oxford University Press.

Mattila, A. S. (2001). Emotional bonding and restaurant loyalty. *Cornell Hotel and Restaurant Administration Quarterly, 42*, 73-79.

Morris, J. A., & Feldman, D. C. (1996). The dimensions, antecedents, and consequences of emotional labor. *Academy of Management Review, 9*, 257-274.

Ostroff, C. (1993). The effects of climate and personal influences on individual behavior and attitudes in organizations. *Organizational Behavior and Human Decision Process, 56*, 56-90.

Ozcelik, H., Langton, N., & Aldrich, H. (2001). Does intention to create a positive emotional climate matter? A look at revenue, strategic and outcome growth. *Academy of Management Best Paper Proceedings,* Washington, D.C.

Peccei, R., & Rosenthal, P. (2000). Front-line responses to customer orientation programs: A theoretical and empirical analysis. *International Journal of Human Resource Management, 11*, 562-590.

Pugh, S. D. (2001). Service with a smile: Emotional contagion in the service encounter. *Academy of Management Journal, 44*, 1018-1027.

Pugh, S. D., Dietz, J. D., Wiley, J. W., & Brooks, S. M. (2002). Driving service effectiveness through employee-customer linkages. *Academy of Management Executive, 16*(4), 73-84.

Rafaeli, A., & Sutton, R. I. (1987). Expression of emotion as part of the work role. *Academy of Management Review, 12*, 23-37.

Rafaeli, A., & Sutton, R. I. (1990). Busy stores and demanding customers: How do they affect the display of positive emotion? *Academy of Management Journal, 33*, 623-637.

Rothbard, N. P. (2001). Enriching or depleting? The dynamics of engagement in work and family roles. *Administrative Science Quarterly, 46*, 655-684.

Rothbard, N., & Wilk, S. L. (2004). *Spillover and contagion: Mood, worker performance, and burnout.* Paper presented at the Academy of management annual meeting, New Orleans, LA.

Ryan, A. M., & Ployhart, R. E. (2003). Customer service behavior. In W. C. Borman. D. R. Ilgen. & R. J. Klimaski (Eds.), *Handbook of Psychology: Industrial and Organizational Psychology* (Vol. 12, pp. 377-397). New Jersey: John Wiley & Sons.

Ryan, A. M., Schmit, M. J., & Johnson, R. (1996). Attitudes and effectiveness: Examining relations at an organizational level. *Personnel Psychology, 49*, 853-882.

Schaubroeck, J., & Jones, J. R. (2000). Antecedents of workplace emotional labor dimensions and moderators of their effects on physical symptoms. *Journal of Organizational Behavior, 21*, 163-183.

Schneider, B., & Bowen, D. E. (1993). The service organization: Human resources management is crucial. *Organizational Dynamics, 21*, 39-52.

Schneider, B., Bowen, D. E., Ehrhart, M. G., & Holcombe, K. M. (2000). The climate for service: Evolution of a construct. In N. M. Ashkanasy, C. P. M. Wilderom, & M. F. Peterson (Eds.), *Handbook of organizational culture and climate* (pp. 21-36). Thousand Oaks, CA: Sage.

Schneider, B., Ehrhart, M. G., Mayer, D. M., Saltz, J. L., & Niles-Jolly, K. (in press). Understanding organizational-customer links in service settings. *Academy of Management Journal.*

Schneider, B., Gunnarson, S. K., & Niles-Jolly, K. (1994). Creating the climate and culture of success. *Organizational Dynamics, 23*, 17-29.

Schneider, B., Wheeler, J. K., & Cox, J. F. (1992). A passion for service: Using content analysis to explicate service climate themes. *Journal of Applied Psychology, 77*, 705-716.

Schneider, B., White, S. S., & Paul. M. C. (1998). Linking service climate and customer perceptions of service quality: Test of a causal model. *Journal of Applied Psychology, 83*, 150-163.

Seo, M. G., Feldman Barrett, L., & Bartunek, J. M. (2004). The role of affective experience in work motivation. *Academy of Management Review, 29*, 423-439.

Staw, B. M., & Barsade, S. G. (1993). Affect and managerial performance: A test of the sadder-but-wiser vs. happier-and-smarter hypotheses. *Administrative Science Quarterly, 38*, 304-331.

Staw, B. M., Sutton, R. R., & Pelled, L. H. (1994). Employee positive emotion and favorable outcomes at the workplace. *Organization Science, 5*, 51-71.

Susskind, A. M., Kacmar, K. M., & Borchgrevink, C. P. (2003). Customer service providers' attitudes relating to customer service and customer satisfaction in the customer-service exchange. *Journal of Applied Psychology, 88*, 179-187.

Tan, H. H., Foo, M. D., & Kwek, M.H. (2004). The effects of customer personality traits on the display of positive emotions. *Academy of Management Journal, 47*, 287-296.

Thoits, P. A. (1996). Managing the emotions of others. *Symbolic Interaction, 19*, 85-109.

Tsai, W. C., & Huang, Y. M. (2002). Mechanisms linking employee affective delivery and customer behavioral intentions. *Journal of Applied Psychology, 87*, 1001-1008.

Tschan, F., Semmer, N. K., & Inversin, L. (2004). Work related and "private" social interactions at work. *Social Indicators Research, 67*, 145-182.

van Dolen, W., de Ruyter, K., & Lemmink, J. (2004). An empirical assessment of the influence of customer emotions and contact employee performance on encounter and relationship satisfaction. *Journal of Business Research, 57*, 437-444.

Walsh, K. (2000). A service Conundrum: Can outstanding service be too good? *Cornell Hotel and Restaurant Administration Quarterly, 41*, 40-50.

Watson, D. (2000). *Mood and temperament.* New York: Guilford Press.

Weiss, H. M., & Cropanzano, R. (1996). Affective events theory: A theoretical discussion of the structure, causes, and consequences of affective experiences at work. *Research in Organizational Behavior, 18*, 1-74.

Wharton, A. S., & Erickson, R. J. (1993). Managing emotions on the job and at home: Understanding the consequences of multiple emotional roles. *Academy of Management Review, 18*, 457-486.

Yagil, D., & Gal, I. (2002). The role of organizational service climate in generating control and empowerment among workers and customers. *Journal of Retailing and Consumer Services, 9*, 215-226.

doi:10.1300/J369v09n02_08

Understanding Culinary Arts Workers: Locus of Control, Job Satisfaction, Work Stress and Turnover Intention

Hsu-I Huang

SUMMARY. A study of Taiwan culinary arts workers was undertaken to reveal the relationship among culinary arts workers locus of control, demographic variables, job satisfaction, work stress, and turnover intention. The results exhibited that male culinary arts workers had a higher degree of internal locus of control than female culinary arts workers. Internal locus of control was significantly and positively correlated with employee job satisfaction and their tenure in the kitchens but significantly and negatively correlated with work stress and turnover intention. Furthermore, the cluster analysis classified culinary arts workers into three heterogeneous groups; each group showed different degrees of internal and external locus of control, turnover intention, job satisfaction, and work stress. These results provided an insight in understanding culinary arts workers and lead to suggestions for developing effective human resource strategies. doi:10.1300/J369v09n02_09 *[Article copies available for a fee from The Haworth Document Delivery Service: 1-800-HAWORTH.*

Hsu-I Huang, EdD, is Assistant Professor, Dept. of Leisure, Recreation, and Tourism Management, Southern Taiwan University of Technology, 29 Chien Fong Rd., Gang Shan, Kaohsiung County, Taiwan, 82046 (E-mail: 8sunny@gmail.com).

[Haworth co-indexing entry note]: "Understanding Culinary Arts Workers: Locus of Control, Job Satisfaction, Work Stress and Turnover Intention." Huang, Hsu-I. Co-published simultaneously in *Journal of Foodservice Business Research* (The Haworth Hospitality & Tourism Press, an imprint of The Haworth Press, Inc.) Vol. 9, No. 2/3, 2006, pp. 151-168; and: *Human Resources in the Foodservice Industry: Organizational Behavior Management Approaches* (ed: Dennis Reynolds, and Karthik Namasivayam) The Haworth Hospitality & Tourism Press, an imprint of The Haworth Press, 2006, pp. 151-168. Single or multiple copies of this article are available for a fee from The Haworth Document Delivery Service [1-800-HAWORTH, 9:00 a.m. - 5:00 p.m. (EST). E-mail address: docdelivery@haworthpress.com].

Available online at http://jfbr.haworthpress.com

doi:10.1300/J369v09n02_09

KEYWORDS. Locus of control, job satisfaction, work stress, turnover intention

INTRODUCTION

Taiwan is becoming a country more oriented towards tourism and leisure businesses. These transformations resulted from the new two-day weekends policy that began from 1996, the emigration of major factories to lower labor cost countries, especially mainland China, and the congenital advantage of abundant ecological resources (e.g., lofty mountains, hot springs, lakes, rivers and creeks) and diverse cultural resources (e.g., historical destinations, folk arts, festival celebrations, and regional delicacies).

These transformations have advanced the development of the international tourist market in Taiwan, especially with the ongoing development of hotels. According to the Taiwan Tourism Bureau (2006), there are four levels of hotels in Taiwan, which include (a) 60 international tourist (the highest level) hotels, with an additional 35 under construction, (b) 27 tourist hotels, (c) 3,253 general hotels, and (d) 801 legal and registered host families. The international tourists have brought economic benefits to Taiwan. It can be seen from a total income of approximately $ 4,200,000,000 U.S. dollars from about 3,000,000 international tourists in 2002 (Taiwan External Trade Council, 2006). In order to reach the goal of attracting 5,000,000 international tourists to Taiwan annually by year 2008, the government of Taiwan is investing $ 2,400,000,000 U.S. dollars in developing new attractive tourist destinations, increasing tourist markets, promoting tourism businesses, and encouraging tourism related investigations (Taiwan External Trade Council, 2006). These issues highlight the importance of the tourism business in Taiwan.

Among the three major reasons attracting international tourists to Taiwan, its world renowned cuisine tops the list (Taiwan Tourism Bureau, 2005). The annual revenue from food and beverage from international tourist hotels achieved 44.9 % of the total revenue of international tourist hotels (Taiwan Tourism Bureau, 2006). Culinary arts workers primarily serve as corner stones in restaurant operations. It would not be possible to impressively serve this cuisine without excellent culinary arts workers. However, due to heavy work loads, rush hour pressure, and work shifts, the turnover rate for hospitality employees in Taiwan typically is high (Huang, 2003). High employee turnover not only leads to high operation costs but also reduces productivity and service quality (Lungberg & Armatas, 1980).

Many research studies have been conducted in the hospitality industry to help employers and researchers understand the relationship between restaurant and hotel employee personal characteristics and job attitude, such as the correlations of employee job satisfaction, turnover intention, work stress (Delton & Todor, 1987; Lee & Mitchell, 1994; McCormic & Ilgen, 1985; Seashore, Lawler, Mirvis, & Cammann, 1983; White & Bednar, 1991). Nevertheless, little study has been concentrated on the culinary arts workers. Thus, the particular purpose of this study is to understand the culinary arts workers in Taiwan by revealing the relationship among their loci of control, job satisfaction, work stress and turnover intention. The significance of the results might provide references for restaurant managers in understanding, recruiting, training and managing the culinary arts workers with effective strategies and thereby enhancing their job satisfaction and performance and reducing their turnover intention.

THEORETICAL BACKGROUND

Several general perspectives on job attitudes suggest that relations exist among job satisfaction, work stress, turnover intention and employee's distinctive characteristics. First, research has demonstrated that high job satisfaction has a beneficial impact on both individuals and organizations where it has comparatively negative correlations with employees work stress and turnover intention (Boxx, Odom, & Dunn, 1991; Lee & Mitchell, 1994; McCormic & Ilgen, 1985; Verquer, Beehr, & Wagner, 2003; Seashore et al., 1983). Second, research on work stress (e.g., Brief, Schuler, & Sell, 1981; Schmitz, Neumann, & Oppermann, 2000) suggests that work stress perception may differ depending on the individual's different degree of internal / external locus of control (Brissett & Nowicki, 1976; Grannis, 1992; Schmitz, Neumann, & Oppermann, 2000) and different personal characteristics (Brief, Schuler, & Sell, 1981; Cartwright & Cooper, 1997). Finally, work stress was revealed to have a positive correlation with turnover intention (Huang, 2005; Mowday, Porter, Steers, 1982). These views converge to suggest different degrees of associations might exist among personal characteristics, job satisfaction, work stress, and turnover intention as depicted in the literature review.

LITERATURE REVIEW

Locus of Control

The theory of locus of control was initially developed by Rotter in 1954. Rotter (1966) proposed that locus of control can be categorized into internal locus of control and external locus of control. People with strong external locus of control tend to believe their successes, failures, good or bad fortune, consequences or results in their lives are controlled by the outside environment, instead of themselves. In contrast, people with strong internal locus of control believe they themselves are in control of reaching their goals, successes, failures or any consequences in the future.

Locus of control was found to be related to work behavior (Spector, 1982); furthermore, it has also been used in 9,339 psychological research studies from 1967 to 1999 (Judge & Bono, 2001). Many researchers have found that people with different loci of control have different attitudes toward their jobs. In comparing external locus of control with internal locus of control, those who are predominantly internal locus of controllers are more satisfied with their pay (Andrasani & Nestel, 1976), jobs (Organ & Greene, 1974), appreciate achievement, have an intensified focus on their personal interests or valued issues, and perform better (Cheng, 1994; Pilling, Donthu, & Henson, 1999; & Kern, 1992). They also are able to withstand more stress than strong external locus of controllers (Brissett & Nowicki, 1976; Grannis, 1992). Thus, these consequences imply that individuals with different loci of control might have different attitudes towards their jobs. Besides loci of control, job satisfaction was also found to have an impact on employee job attitudes.

Job Satisfaction

Research on job satisfaction and its causal correlations to other components of job attitudes have been widespread. Research has manifested that high job satisfaction had a beneficial impact on both individuals and organizations, and it had comparatively negative correlations with employees work stress, absenteeism, and voluntary turnover intention

(Seashore, Lawler, Mirvis, & Cammann, 1983). In addition, job satisfaction was revealed to have positive correlations with employee performance, productivity, and job involvement (Brief, 1998; Ghiselli, La Lopa, & Bai, 2001; Huang, 2003; McCormick & Ilgen, 1985; Milkovich & Boudreau, 1988; Mowday, Porter, & Steers, 1982; Statt, 1994; White & Bednar, 1991). When employees are dissatisfied with their jobs, the consequence of their attitudes may lead employees to withdraw from their jobs and look for other jobs or negatively express their dissatisfaction through absenteeism, tardiness, reducing their efforts, or increasing their error rate (Robbins, 1996). Therefore, this corroborating evidence suggests that there are correlations between employee job satisfaction, turnover intention, work stress and related work attitudes. How work stress affects employee work behavior and how employees handle work stress is another concern that diversifies from individual to individual and affects job attitudes.

Work Stress

Stress had been defined as "external forces," "outside stimulus," and "pressure" on an individual (Cartwright & Cooper, 1997). Major sources of work stress have been categorized into three types: organizational characteristics and process, job demands and role characteristics, and individual characteristics and expectations (Brief, Schuler & Sell, 1981). More specifically, factors such as rotating work shifts, pay inequities, low participation in decision making, poor communication, inadequate training programs, poor work conditions, crowding, noise, excessive heat or cold, light, safety hazards, air pollution, poor interpersonal relationships, time pressures, role conflicts, overload, and personal career concerns also result in work stress (Brief et al., 1981). Culinary arts workers work under the pressure of intense deadlines in an environment filled with heat, smoke, and high humidity, and complicated by a physically demanding work load. Thus, it might be assumed that there are correlations between the uncomfortable environment in the kitchen and employee work stress.

Personality type differences also affect how individuals cope with work stress. A type A person personality, with a desire for control, impatience, aggressiveness, hostility, and irritation, normally exhibit high work stress (Brief, Schuler & Sell, 1981). Cartwright and Cooper (1997) also noted that different personalities may lead some to overcome stress successfully but others to fail. When stress on individuals exceeds their capabilities to overcome it, the stress will continue to exist; On the other hand, successfully coping with stress may lead to personal growth (Williams & Cooper, 2002).

Many research results demonstrated that stress has a strong impact on organizations (Brief et al., 1981; Sauter & Murphy, 1995; White & Bednar, 1991; Williams & Cooper, 2002). The impacts of work stress on organizations, whether visible or invisible, causes a substantial decrease in productivity and an increase in absenteeism as well as a strain on labor and administrators, job dissatisfaction, and turnover intention (Brief et al., 1981; Lingard, 2003; Pinder, 1984; Rasch & Harrell, 1989; Sauter & Murphy, 1995; White & Bednar, 1991; Williams & Cooper, 2002).

Turnover Intention

Employee turnover can be categorized into voluntary turnover (such as quitting a job) and involuntary turnover (such as being laid off). The potential expenditure of restaurant and hotel employee turnover was previously reported to range from $ 50 U.S dollars to thousands of U.S. dollars per individual, depending on the position (Tecipi & Bartlett, 2002).

When employees encounter unsatisfying experiences, these might kindle the intention of leaving (Mobley, 1983). Turnover intention gradually develops before job resignation (Porter & Steers, 1973) and generate in several stages, which include: comprising withdrawing behavior (such as being sick, late, or absent), looking for other job opportunities, evaluating and comparing other jobs, and deciding to quit (Mobley, 1983; Mobley, Horner & Hollingsworth, 1978; Price & Mueller, 1981). The degree of turnover intention was the most important factor in predicting employee leaving (Kraut, 1992; Mobley et al., 1978; Newman, 1974; and Michaels & Spector, 1982).

In addition, some research studies exposed that turnover intention had a strong positive correlation with work dissatisfaction (Dalton & Todor, 1987; Robbins, 1996; Seashore et al., 1983), scarcity of job attraction (Mobley et al., 1978), tenure and age (Huang, 2005, Lambert, Hogan, & Barton, 2001; Mobley et al., 1978), and work stress (Huang, 2005). Huang also found that certain personality traits showed strong negative and positive correlations with hospitality employee turnover intention. Employees with strengths in Sensing, Judging and Thinking personality traits showed a significant and negative correlation with turnover intention whereas employees with strengths in Perceiving and

Feeling personality traits reported a significant and positive correlation with turnover intention (Huang, 2005).

Culinary art workers, as a group, work under time pressure, poor working conditions, and low pay (Huang, 2005); thus, this not only causes high turnover rates, but also affects their attitudes and quality of their work. However, little research has been conducted to explore the relationships among culinary arts workers loci of control, their perceptions of work stress, job satisfaction, and turnover intention. To fill this gap, the current study has been accomplished. The research methodology is presented next.

METHODOLOGY

The purpose of this research was to investigate the relations among culinary arts workers locus of control, job satisfaction, work stress and turnover intention in the international tourist (highest level) hotels in Taiwan. Since there are more restaurants and more professional culinary arts workers in the international tourist hotels than in other levels of hotels, this research focused on the culinary arts workers in the international tourist hotels. This research was conducted to answer the following research questions.

Research Questions

Research question 1. Does locus of control differ in culinary arts workers depending on age, gender, number of years employed in the kitchens, years of employment with current employers, worked kitchens, educational level and monthly salary?

Research question 2. Are there any significant correlations among employee different loci of control, job satisfaction, work stress and turnover intention?

Research question 3. What is the typology of culinary arts workers regarding their different degrees of external and internal loci of control, job satisfaction, educational levels, monthly salary, age, number of years employed in the kitchens and years of employment with current employers?

Samples and Sample Method

A stratified sampling method was used to select participants from the 60 international tourist hotels in Taiwan. The researcher randomly called human resource managers of the 60 international tourist hotels by geographical location in each of the four geographical regions (northern, southern, mid-western, and eastern). Then, the researcher explained the purpose of this research and asked for permission to use their employees and their help in distributing the questionnaires. When the number of hotels reached the number of required samples for each geographical region, the researcher stopped and repeated the same procedure in a different geographical region. A total of 500 questionnaires were randomly distributed to 11 international tourist hotels by human resource managers to culinary arts workers in their different kitchens.

Survey Instruments

The survey instrument consisted of 12 demographic questions; 20 questions about locus of control, modified from those of Rotter (1966) and Spector (1982); 17 work stress perception questions; 15 job satisfaction questions; and 8 questions regarding turnover intention questions modified from the Michigan Organizational Assessment Questionnaire (Seashore, Lawler, Mirvis, & Cammann, 1983) and from the Measurement of Organizational Commitment (Mowday, Steers, & Porter, 1979). All the questions, except demographic ones, were designed with 5-point Likert scales, with "1" corresponding to "strongly agree" and "5" corresponding to "strongly disagree."

Validity and Reliability

Before formal distribution, the survey instrument was reviewed by a panel of consultants, consisting of chefs from the 11 selected international tourist (highest level) hotels, and was revised based on their suggestions. Then, a pilot study was conducted with 30 volunteer chefs and cooks from the participating hotels. Further changes were made to attain content validity and reliability. The reliability of Cronbach's Alpha coefficients was .79 for locus of control, .90 for job satisfaction, .74 for turnover intention, and .91 for work stress.

Data Collection and Data Analysis

The data collection instruments were distributed via mail to human resource managers who had agreed to voluntarily engage in this study. Of the 500 questionnaires distributed, 240 questionnaires were usable, an actual return rate of 49.8%. The data was analyzed by SPSS 11 and statistical methods of descriptive analysis, Pearson bivariate correlation, independent sample T-test, Analysis of Variance (ANOVA) and cluster analysis were applied.

RESULTS

Respondent Profile

The overall gender distribution of the participants in this study was 208 (86.7%) males and 32 (13.3) females. Martial status was 106 (44.2%) married and 134 (55.8%) single. The age of most ranged from 20 to 30 years old (n = 109, 45.4%) followed by 31 to 40 years old (n = 74, 30.8%). The accumulated number of these two groups was 183 (76.2%). Further, 36 (15%) of the culinary arts workers held college degrees, 124 (51.7%) were high school or vocational school graduates, 61 (25.4%) were junior college graduates, 18 (7.5%) were elementary school graduates and there was only 1 (0.4%) respondent who held a master's degree. There were 107 (44.6%) who had graduated with related professional hospitality related backgrounds, which included 53 (22.1%) from Chinese Culinary Arts, 14 (5.8%) from Western Culinary Arts, 5 (2.1%) from Baking and Pastries, 4 (1.7%) from Home Economics, 23 (9.6%) from Restaurant Management, and 8 (3.0%) from Tourism Management.

Among the respondents, 118 (49.2%) worked in Chinese cuisine kitchens, 67 (27.9%) worked in western cuisine kitchens, 11 (4.6%) worked in Japanese cuisine kitchens, 28 (11.7%) worked in baking and pastry kitchens and 16 (6.7%) worked in other kitchens. The majority (69.5%) of their average monthly salary ranged between $17,000 to $45,000 N.T. ($ 531 to 1,406 U.S.). All of the respondents are full time culinary workers, 109 (45.4%) of them worked 8 hours per day and 117 (48.8%) of them worked over 8 hours to 10 hours per day. The most frequently reported period of employment with the current employer was under 1 year (n = 78, 32.5 %), followed by 2 years to 4 years (n = 56, 23.3%) and over 1 year to 2 years (n = 56, 23.3%). In terms of professional experience in kitchens, the most frequently reported period was over 10 years (n = 83, 34.6%) and only 28 (11.7%) of them had less than one year of working experience in kitchen (see Table 1). This result demonstrated that 34.6% of the culinary arts workers have worked in kitchens over 10 years, but surprisingly, 47.1% of them have worked under 1 year to 2 years for their current employers (see Table 1). It suggested that the high employee turnover problems existed among these culinary arts workers.

TABLE 1. Years of Employment with the Current Employers vs. Number of Years Employed in Kitchens

Years	Years of Employment with Current Employers		Number of Years Employed in Kitchens	
	Frequency	Percent	Frequency	Percent
Under 1 Year	78	32.5	28	11.7
Over 1 Year to 2 Years	35	14.6	23	9.6
Over 2 Years to 4 Years	56	23.3	24	10
Over 4 Years to 6 Years	24	10	25	10.4
Over 6 Years to 8 Years	25	10.4	25	10.4
Over 8 Years to 10 Years	8	3.3	32	13.3
Over 10 Years	14	5.8	83	34.6
Total	240	100	240	100

The Results of ANOVA and Post Hoc Analyses

To answer research question 1, Levene's test was first applied to test the assumption of homogeneity of variance among the internal and external locus of control groups. Next, one-way ANOVA was applied to analyze whether there were significant differences in employee job satisfaction, work stress, and turnover intention between the two different loci of control. Following this, a Sheffé post hoc test was performed to compare the differences among the groups. Levene's test of homogeneity of variance indicated all significant values for the above variables exceeded 0.05; therefore, the variances for the groups were equal and the assumption was justified.

The two-tailed independent samples T-test and the ANOVA results showed first, male culinary arts workers had significant higher internal locus of control (n = 206, $M = 4.06$, $p = 0.01 < 0.05$) than female culinary arts workers (n = 32, $M = 3.75$, $p = 0.01 < 0.05$). Internal locus of control differed significantly by educational majors ($p = 0.007 < 0.005$), monthly salary ($p = 0.001 < 0.05$), number of years employed in the kitchens ($p = 0.00 < 0.05$) and years of employment with current employers ($p = 0.044 < 0.05$) but marital status ($p = 0.203 > 0.05$), age ($p = 0.144 > 0.05$), and educational levels ($p = 0.17 > 0.05$) showed no significant differences. The mean scores of internal locus control among the significant variables are displayed in Table 2.

TABLE 2. Mean Scores of Internal Locus of Control Among Demographical Variables

	Mean	Rank * Highest † Lowest	N	Std. Deviation
Monthly Salary N.T. dollars				
Under 17,000	10.84	†	19	2.587
17,001 ~ 25,999	11.69		59	2.027
26,000 ~ 35,999	12.30		59	1.793
36,000 ~ 45,999	12.22		45	1.730
46,000 ~ 55,999	12.10		37	1.696
56,000 ~ 65,999	13.75		4	1.258
66,000 ~ 75,999	13.00		9	1.581
76,000 ~ 85,999	12.80		5	1.09
86,000 ~ 95,999	**15.00**	*	1	.
Number of Years Employed in Kitchens				
Under 1	10.55	†	28	1.77
Over 1 to 2	11		23	2.58
Over 2 to 4	**12.87**	*	23	1.84
Over 4 to 6	12.48		25	1.64
Over 6 to 8	12.44		25	1.89
Over 8 to 10	11.83		31	1.81
Over 10 to 15	12.48		31	1.18
Over 15	12.5		52	1.74
Years of Employment with Current Employers				
Under 1	11.82		78	2.03
Over 1 to 2	11.58	†	34	2.36
Over 2 to 4	**12.69**	*	55	1.82
Over 4 to 6	11.91		24	1.71
Over 6 to 8	12.32		25	1.57
Over 8 to 10	12		8	1.41
Over 10	11.93		14	1.33
Educational Major				
Chinese Culinary Arts	11.5		53	2.20
Western Culinary Arts	10.54	†	13	1.71
Baking & Pastries	12.2		5	0.83
Home Economics	11.5		4	1
Restaurant management	12.22		23	1.56
Tourism Management	12.13		8	2.64
Others	12.42	*	132	1.79

Note. $ 1 U.S. dollar approximately equals to $ 32 N.T. dollar.

The Results of Pearson Bivariate Correlation Analysis

The results revealed that culinary arts workers with strong internal locus of control were significantly and negatively correlated with work stress (–.44, $p = 0.00 < 0.05$), turnover intention (–.309, $p = 0.00 < 0.05$) but significantly and positively correlated with job satisfaction (.47, $p = 0.00 < 0.05$). In contrast, culinary arts workers with strong external locus of control showed a significant and positive correlation with work stress (.26, $p = 0.00 < 0.05$) and turnover intention (.276, $p = 0.00 < 0.05$) and significant and negative correlation with job satisfaction (–.259, $p = 0.00 < 0.05$). Furthermore, internal locus of control had a significant and positive correlation with tenure. These results answered research question 2; if there are significant correlations among locus of control, work stress, job satisfaction, and turnover intention (see Table 3).

The Results of Cluster Analysis

Before performing the cluster analysis, all variables including internal locus of control, external locus of control, job satisfaction, work stress, turnover intention, number of years employed in kitchens, monthly salary,

TABLE 3. Correlations of Internal Locus of Control, External Locus of Control, Work Stress, Job Satisfaction and Turnover Intention

		Internal Locus of Control	External Locus of Control	Work Stress	Job Satisfaction	Turnover Intention
Internal Locus of Control	Pearson Correlation	1.000	–.213**	–.387*	.483**	–.211*
	Sig. (2-tailed)	.	.001	.000	.000	.001
External Locus of Control	Pearson Correlation	–.213	1.000	.362**	–.305**	.153*
	Sig. (2-tailed)	.001	.	.000	.000	.019
Work Stress	Pearson Correlation	–.387**	.362**	1.000	–.542**	.502*
	Sig. (2-tailed)	.000	.000	.	.000	.000
Job Satisfaction	Pearson Correlation	.483**	–.305**	–.542**	1.000	–.412*
	Sig. (2-tailed)	.000	.000	.000	.	.000
Turnover Intention	Pearson Correlation	–.211**	.153*	.502**	–.412**	1.000
	Sig. (2-tailed)	.001	.019	.000	.000	.

** Correlation is significant at the 0.01 level (2-tailed).
* Correlation is significant at the 0.05 level (2-tailed).

years of employment with current employers, age, education were standardized. A hierarchical cluster analysis using Ward's minimum variance method was performed first and a k-mean cluster analysis was conducted by examining the outliers and removing the outliers. The optimum number of clusters was chosen based on the centroid estimate results and the dendrograms. All variables were significant in the ANOVA results of the cluster analysis. The consequences of the final cluster analysis categorized the culinary arts workers into three groups with high internal homogeneity and high external heterogeneity.

Group 1, with the highest level of internal locus of control, had the highest level of job satisfaction and the lowest level of work stress and turnover intention. The age range of this group fell to the middle among these three groups (see Table 4). This group of culinary arts workers had the lowest educational level, but had the highest monthly salaries, longest work experience in kitchens and had an average of 2 to 4 years employment with their current employers. Group 2, with moderate internal

TABLE 4. Comparison of Three Cluster's Characteristics Regarding Employees Demographics, Job Satisfaction, Work Stress, Turnover Intention, and Internal/External Locus of Control

	Cluster Means		
	Group 1 High internal locus of control, high job satisfaction, low work stress & turnover intention group	**Group 2 Moderate Internal Locus Control group**	**Group 3 High external locus control, high turnover intention and work stress group**
1. Age	.06687	.79992	−.69251
2. Education	−.15632	−.53315	.55548
3. Monthly Salary	.55307	.43378	−.76646
4. Number of Years Employed in Kitchens	.50641	.71757	−.96320
5. Years of Employment with Current Employers	−.18959	.93057	−.57746
6. Internal Locus of Control	.74099	−.26766	−.31438
7. External Locus of Control	−.66785	.12090	.36455
8. Turnover Intention	−.83829	.17783	.43345
9. Work Stress	−.95085	.15190	.58052
10. Job Satisfaction	.96454	−.46724	−.32871

and locus of control, had the lowest mean score of job satisfaction, and the highest mean scores of work stress and turnover intention. The age range of this group was the highest one, had the second highest monthly salaries, the second highest amount of years employed in kitchens, and whose average tenure in current employers was between 4 to 6 years. Group 3, the highest external locus of control group, had the highest work stress, moderate levels of turnover intention and job satisfaction, earned the lowest monthly salaries, had the least work experience in the kitchens and least length of tenure in current employers. These outcomes imply that culinary art workers with strong internal locus of control had the highest level of job satisfaction and the lowest levels of work stress and turnover intention (See Table 4). The high turnover intention group tended to be the youngest group, with the highest educational level, the highest mean score on external locus of control, and the least amount of work experience in kitchens and years of employment with current employers.

CONCLUSIONS

The results of this research exposed several different correlations among culinary arts workers loci of control, job satisfaction, work stress, turnover intention, and demographic information. First, from the results of the correlation analysis, culinary arts workers with strong internal locus of control reflected significant and positive correlations with job satisfaction, and number of years employed in kitchens. In contrast, employees with strong external locus of control had significant and negative correlations with job satisfaction, number of years employed in kitchens, and work stress but had a significant and positive correlation with turnover intention. Second, when examining the cluster information, each clustered group also showed different degrees of internal and external locus of control, job satisfaction, work stress, and turnover intention. These results echoed previous research (Brissett & Nowicki, 1976; Grannis, 1992; Huang, 2003; Schmitz, Neumann, & Oppermann, 2000) described in the literature review. Therefore, the results support that employee's with different loci of control has an effect on job satisfaction, work stress, and turnover intention.

The results from ANOVA and the Sheffé post hoc test manifested that loci of control differed according to the demographics of culinary arts workers. Furthermore, job satisfaction and work stress were found to have significant differences depending on employee demographics.

RECOMMENDATIONS

The results of this research provide a beneficial insight to better understanding culinary arts workers regarding their locus of control, job satisfaction, work stress and turnover intention. Since the average number of years worked in kitchens is from two to four years, inferring the factors which affecting this are critical if management wishes to prolong worker tenure. It may also aid in more effective recruiting efforts. This research also showed that the average salary of culinary arts workers was low considering their lengthy work hours and workloads. It is also important for hotel and restaurant management to be aware of this and drastically improve the low salaries and work loads. Doing so might improve culinary arts workers' job satisfaction and encourage them to stay on their job longer. Further, from the information of the three classified culinary arts groups, it is surprising to see the youngest, most educated group having the highest external locus of control and the least years of work experience in kitchens as well as years of employment with current employers exhibited the highest level of turnover intention and work stress. It seems that this group of culinary arts workers tends to be new entry employees. This result might give scholars and hotel and restaurant owners an awareness of further understanding what this group of culinary arts workers need and how to provide suitable environment, benefit, or training programs to satisfy and motivate them.

It is important to address the limitations of this study and these might be investigated in future research. First, this research only concentrated on some of the factors related to turnover intention, influencing factors such as individual's ability, work value, personal/family reasons, better pay elsewhere, or opportunity of finding another job might also influence their turnover intention. Therefore, further research might embrace these factors when possible. Second, this study sampled only 11 five-star hotels in Taiwan; thus, the results may not be representative of all the different types of hotels in Taiwan. Further research could be done in other kinds of hotels and restaurants. This might reveal if the outcomes will be generated differently. This could aid management in recruiting, selection, training, and supervising culinary arts workers with different characteristics. Last, further insights might be acquired through qualitative research. For instance, researchers might gain beneficial information through structured exit interviews, such as finding the reasons why culinary arts employees withdraw from their jobs and what strategies might reduce their work stress, turnover intention and enhance job satisfaction.

REFERENCES

Andrasani, P. J. and G. Nestel (1976), Internal-external control as contributor to and outcome of work experience, *Journal of Applied Psychology*, 156-165.

Boxx, W. R., Odom, R. Y., & Dunn, M. G. (1991). Organizational values and value congruency and their impact on satisfaction, commitment, and cohesion: An empirical examination within the public sector. *Public Personnel Management, 20*(1), 195-205.

Brief, A. P. (1998). *Attitudes in and around organizations*. Thousand Oaks, CA: Sage Publications.

Brief, A. P., Schuler, R. S., & Sell, M. V. (1981). *Managing job stress*. Boston, MA: Little, Brown and Company. .

Brissett, M., & Nowicki, Jr. S. (1976), Internal versus external of reinforcement and reaction to frustration. *Journal of Abnormal and Social Psychology*, (25), 35-39.

Cartwright, S., & Cooper, C. L. (1997). *Managing workplace stress*. Thousand Oaks, CA: Sage Publications.

Cheng, Y. C. (1994). Locus of control as an indicator of Hong Kong teachers' job attitudes and perceptions of organizational characteristics. *Journal of Education Research, 87*(3), 180-188.

Dalton, D. R., & Todor, W. D. (1987). Turnover turned over: An expanded and positive perspective. In W. O. Dennis (Ed.), *The applied psychology of work behavior* (3rd ed., pp. 253-270). Plano, TX: Business Publication, INC.

Ghiselli, R. F., La Lopa, J. M., & Bai, B. (2001). Job satisfaction, life satisfaction and turnover intent. *Cornell hotel and restaurant administration quarterly* (April), 28-37.

Grannis, J. C. (1992). "Students stress, distress, and achievement in an urban intermediate school," *Journal of Early Adolescence*, (February), 4-27.

Huang, H. I. (2003). *Investigation of the relationships of employee locus of control, job satisfaction, work stress and turnover intention*. The 3rd Annual Tourism, Leisure and Hospitality Management Conference Proceedings, 441-454, Kaohsiung, Taiwan.

Huang, H. I. (2005). *Investigation of the fit among current and preferred organizational cultures, personality styles, and job attitudes in employees of international tourist hotels in Taiwan*, Unpublished Dissertation, Idaho State University, Pocatello, Idaho.

Judge, T. A. & Bono, J. E. (2001). "Relationship of core self-evaluation, self-esteem, generalized self-efficacy, locus of control, and emotional stability-with job satisfaction and job performance: A meta-analysis." *Journal of Applied Psychology, 86*(1), 80-92.

Kern, L. (1992), The moderating effects of locus of control on performance incentives and participation. *Human Relations, 45*(9), 991-1012.

Kraut, A. I., (1992). Predicting turnover of employees from measured job apt. *Academy of Management Journal, 23*, 3-5.

Lambert, E. G., Hogan, N. L., & Barton, H. M. (2001). The impact of job satisfaction on turnover intent: a test of a structural measurement model using a national sample of workers. *The Social Science Journal* (38), 233-255.

Lee, T. W., & Mitchell, T. R. (1994). Organizational attachment: Attitudes and actions. In J. Greenberg (Ed.), *Organizational Behavior* (pp. 83-108). Hillsdale, New Jersey: Lawrence Erlbaum Associates.

Lingard, H. (2003). The impact of individual and job characteristics on 'burnout' among civil engineers in Australia and the implications for employee turnover. *Construction Management and Economics* (21), 69-80.

Lungberg, D. E., & Armatas, J. P. (1980). *The management of people in hotels, restaurants and club* (4th ed.). Dubuque, Iowa: Wm. C. Brown.

McCormick, E. J., & Ilgen, D. (1985). *Industrial and organizational psychology* (8th ed.). Englewood Cliffs, New Jersey: Prentice-Hall, Inc.

Michaels, C. E., & Spector, P. E. (1982). Causes of employee turnover: A test of the Mobley, Griffeth, Hand, and Meglino model. *Journal of Applied Psychology, 67*(1), 53-59.

Milkovich, G. T., & Boudreau, J. W. (1988). *Personnel human resource management: A diagnostic approach* (5th ed.), Business Publications.

Mobley, W. H., Horner, S. O., & Hollingsworth, A. T. (1978). An evaluation of precursors of hospital employee turnover. *Journal of Applied Psychology, 63*(4), 408-414.

Mobley, W. H. (1983). Intermediate linkages in the relationship between job satisfaction and employee turnover. In B. M. Straw (Ed.), *Psychological foundations of organizational behavior* (2nd ed., pp. 107-110). Glenview, Illinois: Scott, Foresman and Company.

Mowday, R. T., Porter, L. W., & Steers, R. M. (1982). *Employee-organization linkages: The psychology of commitment, absenteeism, and turnover*. New York: Academic Press.

Mowday, R. T., Steers, R. M., & Porter, L. W. (1979). The measurement of organizational commitment. *Journal of Vocational Behavior*, 14, 224-247.

Newman, J. E. (1974). Predicting absenteeism and turnover. A field comparison of fishbein's mold and traditional job attitude measures. *Journal of Applied Psychology, 27*(2), 330-350.

Organ, D., Greene, C. (1974). Role ambiguity, locus of control and work satisfaction. *Journal of Psychology, 59*, 101-102.

Pilling, B. K., Donthu, N., & Henson, S. (1999). "Accounting for the impact of territory characteristics on sales performance: Relative efficiency as a measure of salesperson performance." *The Journal of Personal Selling & Sales Management, 19*(2), 35-45.

Pinder, C. C. (1984). *Work Motivation*. Glenview, IL: Scott, Foresman and Company.

Porter, L. W. & Steers, R. M. (1973). Organizational, work and personal factors in employee turnover and absenteeism. *Psychological Bulletin, 80*(2), 151-176.

Rasch, R. H., & Harrell, A. (1989). The impact of individual differences on MAS personnel satisfaction and turnover intentions. *Journal of Information Systems, Fall*, 13-22.

Robbins, S. P. (1996). *Organizational behavior: Concept, controversies, applications* (7th ed.). Englewood Cliffs, NJ: Prentice Hall.

Rotter, J. B. (1966). Generalized expectancies for internal versus external control of reinforcement. *Psychological Monographs, 80*, 1-28.

Price, J. L., & Mueller, C. W. (1981). A casual model of turnover for nurses. *Academy of Management Journal, 24*(3), 543-556.

Sauter, S. L., & Murphy, L. R. (1995). *Organizational risk factors for job stress.* Washington, DC: American Psychological Association.

Schmitz, N., Neumann, W., Oppermann, R. (2000). Stress, burnout and locus of control in German nurses, *International Journal of Nursing Studies, 37*, 95-99.

Seashore, S. E., Lawler III, E. E., Mirvis, P. H., & Cammann, C. (Eds.). (1983). *Assessing organizational change. A guide to methods, measures, and practices.* New York, NY: John Wiley & Son.

Spector, P. E. (1982). "Behavior in Organizations as a Function of Employee's Locus of Control," *Psychological Bulletin, 91*, 482-497.

Statt, D. A. (1994). *Psychology and the world of work.* Washington Square, New York: New York University Press.

Taiwan External Trade Council (2006). *Investment of Tourism, Leisure, and Recreation Businesses in Taiwan.* (Feb. 2006). Retrieved Feb. 28, 2006. from http://www.investintaiwan.org.tw/theme02_c/c_t02_01_10.htm.

Taiwan Tourism Bureau. (2005). *Monthly report on tourist hotel operations in Taiwan (Jan. to Dec. 2004).* Retrieved Feb. 14, 2005, from http://202.39.225.136/statistics/month2.asp?relno = 96.

Taiwan Tourism Bureau (2006). *Monthly Report on International & Standard Tourist Hotel Operations, (Jan. 2006).* Retrieved Feb. 27, 2006, from http://202.39.225.136/indexc.asp.

Tepeci, M., & Bartlett, A. L. (2002). The hospitality industry culture profile: A measure of individual values, organizational culture, and person-organization fit as predictors of job satisfaction and behavioral intentions. *Hospitality Management* (21), 151-170.

Verquer, M. L., Beehr, T. A., & Wagner, S. H. (2003). A meta-analysis of relations between person-organization fit and work attitudes. *Journal of Vocational Behavior, 63*, 473-489.

White, D. D., & Bednar, D. A. (1991). *Organizational behavior. Understanding and managing people at work* (2nd ed.). Needham Heights, MA: Allyn and Bacon.

Williams, S., & Cooper, L. (2002). *Managing workplace stress. A best practice blueprint.* West Sussex, England: John Wiley & Sons, Ltd.

doi:10.1300/J369v09n02_09

Management Perceptions of Older Employees in the U.S. Quick Service Restaurant Industry

Robin B. DiPietro
Merwyn L. Strate

SUMMARY. The quick service restaurant industry has had a consistent problem finding employees for its growing sales trends over the past decade. The traditional employee for quick service restaurants is often young and inexperienced. The current study is exploratory research into the quick service restaurant industry staffing practices and procedures done by analyzing the perceptions of 20 managers from quick service restaurant chains in the Southeastern part of the U.S. regarding older workers. The study finds that the positive perception held by managers regarding older workers does not translate into a stated desire to actually

Robin B. DiPietro, PhD, is Assistant Professor, Rosen College of Hospitality Management, University of Central Florida, 9907 Universal Blvd., Orlando, FL 32819 (E-mail: dipietro@mail.ucf.edu).

Merwyn L. Strate, PhD, is Assistant Professor, Organizational Leadership and Supervision, Purdue University College of Technology, 2325 Chester Boulevard, Richmond, IN 47374-1289 (E-mail: mstrate@purdue.edu).

[Haworth co-indexing entry note]: "Management Perceptions of Older Employees in the U.S. Quick Service Restaurant Industry." DiPietro, Robin B., and Merwyn L. Strate. Co-published simultaneously in *Journal of Foodservice Business Research* (The Haworth Hospitality & Tourism Press, an imprint of The Haworth Press, Inc.) Vol. 9, No. 2/3, 2006, pp. 169-185; and: *Human Resources in the Foodservice Industry: Organizational Behavior Management Approaches* (ed: Dennis Reynolds, and Karthik Namasivayam) The Haworth Hospitality & Tourism Press, an imprint of The Haworth Press, 2006, pp. 169-185. Single or multiple copies of this article are available for a fee from The Haworth Document Delivery Service [1-800-HAWORTH, 9:00 a.m. - 5:00 p.m. (EST). E-mail address: docdelivery@haworthpress.com].

Available online at http://jfbr.haworthpress.com

doi:10.1300/J369v09n02_10

hire a larger percentage of older workers to staff the restaurants. Implications for practitioners are presented. doi:10.1300/J369v09n02_10 *[Article copies available for a fee from The Haworth Document Delivery Service: 1-800-HAWORTH. E-mail address: <docdelivery@haworthpress.com> Website: <http://www.HaworthPress.com>*

KEYWORDS. Quick service restaurant industry, older workers, employee staffing, management perceptions, hospitality industry

INTRODUCTION

The foodservice industry represents $476 billion in revenue and over 4% of the gross domestic product of the United States (U.S.) (National Restaurant Association, 2005). There are currently more than 12 million workers employed in the U.S. foodservice industry (National Restaurant Association, 2005). This makes the foodservice industry the number one private sector employer in the U.S. by employing over 9 percent of the total population (National Restaurant Association, 2003; Spielberg, 2004). Restaurant industry sales have been steadily increasing for the past 13 years and it is projected that they will continue increasing (National Restaurant Association, 2005). Due to this continuing growth, effective recruiting and retention of hourly employees remain major challenges that restaurant managers must resolve (National Restaurant Association, 2003).

The demographics of the world have changed so significantly that businesses of all types will soon have to rely heavily upon "older" workers to staff their vacant positions. According to the U.S. Department of Labor (2005), in 2003, older workers represented 15.4% of the total workforce. This shift in demographics will require proactive implementation of policies and procedures to attract older workers and ensure that organizational development processes are in place to help with the change in the culture of many organizations as old paradigms related to the hiring of employees have to adjust (Daft, 1998).

The fastest growing segment of the workforce is 55- to 64-year olds. By 2020, 37% of the U.S. population will be over the age of 50 (AARP, 2001). The talent pool is getting significantly older as the baby boomers approach the traditional ages of retirement. There were decades of low birth rates which are now leading to a shortage of younger workers (Burtless, 1998). Younger workers have been the traditional workforce of the quick service restaurant industry (DiPietro & Milman, 2004). With the increasing shortage of available younger workers, businesses in the restaurant industry will simply have no choice but to turn to older workers to staff their vacant positions.

Management perceptions of this soon to be large cadre of older workers are important in driving organizational development practices and change in organizations as top management has to drive the paradigm shift in the organization (Burke, 1994). Learning about older workers and subsequently assimilating them into the workplace is one of the most pressing issues faced by business today (Doverspike, 2000). The aging workforce creates unique challenges to the business world in general and to the restaurant industry specifically.

The current study will determine the perceptions of management regarding the hiring and use of older workers in the quick service restaurant industry. The management perceptions regarding older workers are important to know because they will in turn drive the organizational policies and practices for the future. Organizational development focuses on the development and fulfillment of people in organizations to bring about improved performance (Daft, 1998). Human resource development (HRD) professionals have amassed an impressive body of knowledge on how to assemble, train and sustain a competent workforce, but they face a daunting challenge over the next ten years, as the talent pool gets significantly older (Strate & Torraco, 2005).

BACKGROUND LITERATURE

Quick Service Restaurant Industry

According to the National Restaurant Association, in 2004 the quick service segment of the U.S. restaurant industry will account for almost half of all commercial restaurant "eating place" sales, accounting for revenues in excess of $119 billion (Spielberg, 2004). Quick service restaurants, also known as fast food restaurants, are defined by the U.S. National Adult Tracking Surveys Service (2001) as foodservice establishments that have limited service and menu items. Increasing sales and continued high turnover in this segment of the restaurant industry has made staffing a challenge (National Restaurant Association, 2005).

Despite the fact that there is an increasing need for labor and an increasingly older workforce, older workers have not typically been viewed as the "labor source" for quick service restaurants. The quick service restaurant industry has demographics that reflect a younger and less experienced employee and manager. According to the Bureau of Labor Statistics (2000), less than 8% of employees and managers in foodservice operations are over 55 years of age. The primary age group of foodservice employees is 18-24 years old. The largest age group of managers in foodservice jobs is 35-44 years old (National Restaurant Association, 2003). There is distinct gap between the need for hourly employees and managers, and the recruitment efforts and use of an aging workforce in the U.S. by the foodservice industry.

Employers in the restaurant industry have hired a seemingly inexhaustible supply of younger workers in years past. The supply has now evaporated and baby boomers are just beginning to reach retirement age. Moreover, they will not be retiring en masse but rather over the course of many years. Employers have been complaining for several years about the difficulties they face when it comes to finding and hiring skilled workers (Cheney, 2001).

Older Workers

The effects of the aging workforce have already been felt in the industrial nations of the world. Japan, the U.S., and Western Europe are all rapidly aging societies (Schwartz, 2003). In the U.S., the demographers at the Social Security Administration expect the number of elderly people to double by 2035. Currently, there are five Americans in their prime working years (20 to 64) for every individual over 65. Assuming no massive changes in immigration policy, that ratio will change to three-to-one in 2025. By 2075 the two groups will be virtually the same size (Schwartz, 2003). Because of increased medical knowledge and technology, along with healthy diets and lifestyles, the number of physically able older people will continue to increase (Barth & McNaught, 1991).This radical change in the demographics of the workforce has never been experienced and there is little research to guide organizations in addressing this relevant issue in the workplace.

Haider and Loughran (2001) observe that workers age 65 and older, whom they call "elder" workers, tend to be healthier, better educated, and more affluent than their non-working peers. But the researchers also observe that elder workers are working for relatively low wages. This situation appears to be the result of "selecting into low wages" as an exchange for flexibility in their jobs. If this is true, work may be close to leisure for the majority of older workers, suggesting the need for improving opportunities for older workers returning to work (Haider &

Loughran, 2001). Older workers have a strong desire and/or need to continue working (Bolch, 2000; Magd, 2003). This possibility certainly calls for more research in order to change the current paradigm regarding recruiting and staffing of employees in most businesses (Rix, 2002).

Research done by Barth (2000) and Barth and McNaught (1991) has shown that a significant percentage of older workers not currently working would like to work and that older workers can be as cost effective and capable as younger workers. Despite the increase in the number of older workers available and the continued labor shortage in many industries, there is not much literature found on the aging workforce and HRD policies in organizations in general and in quick service organizations specifically (Herman, Olivo & Gioia, 2003; Rocco, Stein & Lee, 2003).

There have not been many large scale studies done on employers' perceptions of programs and policies for the hiring and use of older workers in the hospitality industry. Nevertheless, the consistency of results across smaller scale studies over the years indicates that certain employer opinions of older workers are widespread and entrenched. Employers regard older workers as mature, experienced, loyal, and dependable. However, they harbor reservations about older workers' technological competence, ability to learn, flexibility, adaptability, and cost. For these and perhaps other reasons, older workers still face formidable and often insurmountable barriers to finding work (Magd, 2003; Rix, 2002).

Past Research on Older Workers

Past attitudes by employers and managers may be a reason why businesses know so little about older workers and why they are so ill equipped to deal with older workers. Myths notwithstanding, the differences between older and younger workers' productivity are small and often can be addressed successfully (Wade, 2003). There are studies that suggest attitudes toward older workers are a cultural issue. In a study by Gibson, Zerbe and Franken (1993), they found that American, as well as most other modern societies, are generally negative in attitudes toward older individuals. The range of management attitudes toward older workers is quite broad. A study by Sargeant (2001) found that 43% of managers believed older workers are hard to train; 22% felt them to lack creativity; 38% believed them too cautious, 40% believed them to have difficulty adapting to new technology; and 27% perceived them to be inflexible. On the other hand, 74% believed them to be more reliable than younger workers (Sargeant, 2001). One researcher described a virtual life cycle of attitudes regarding older workers in which people are not hired after they are 40, not trained after they are 50, and offered incentives to leave by the time they are 55 (Hignite, 2000).

Other negative attitudes and perceptions have indicated that older workers are unable to learn and understand new technologies (AARP, 2000). Older workers were also perceived by management as having income that fell below or just at the poverty line and were lonely (Doering, Rhodes & Schuster, 1983; Hassell, 1991; Tomb, 1984). In a study done by Singer and Sewell (1989), when given the choice between an older or younger candidate for the same job, managers tended to place the older worker in a low-status job. In addition, very few companies have programs or policies specifically designed for full utilization of older workers (AARP, 2000).

Past research has been done by Meyer and Meyer (1988), Tayler and Walker (1993), and Magd (2003) regarding hospitality managers' attitudes and perceptions of older employees in hospitality organizations. These studies have shown that managers perceive that the image of the hotel or hospitality workplace would be negatively impacted by the presence of older workers in the service areas of the hotel, and have therefore kept older employees away from front line positions because of these fears (Meyer & Meyer, 1988). It has been shown through the literature and past studies that most hospitality employers hold some negative perceptions and negative attitudes toward the employment of older workers regarding employees' flexibility, adaptability to change and productivity (Magd, 2003; Meyer & Meyer, 1988; Tayler & Walker, 1993).

DeMicco and Reid (1988) specifically analyzed the use of older workers in the restaurant industry and found distinct attributes regarding the use of older workers in this industry. DeMicco (1989) studied the use and perceptions of the use of older workers in institutional foodservice organizations. Older workers were found to be more dependable, have fewer absences, fewer on the job injuries and were more satisfied with their jobs than their younger counterparts. Older workers also encountered decreased job-related stress, and in the quick service restaurant industry, this would be a very positive trait to have. These studies did not specifically address management's perceptions of the use of older workers in the staffing of quick service restaurant operations, which rely more heavily on a younger workforce.

Studies showed that one of the most important perceived benefits that older workers brought to the company was that they were less accident prone and more cautious than their younger counterparts (Dibden & Hibbert, 1993; Magd, 2003; Thatcher, 1996). Older workers also have lower turnover and take less time off (Perry, 2005). In addition, the

National Council on Aging has found that older workers had a higher level of commitment to their companies than younger workers. Older workers are also more valuable to their companies with respect to work ethic, accuracy and stability (Catrina, 1999; Kaeter, 1995; Stein & Rocco, 2001). In fact, some older workers are able to mentor the younger workers in problem solving (Perry, 2005). Furthermore, it has been found that older workers tend to better relate to and spend time building good rapport with customers than younger workers (*Caterer and Housekeeper*, 1992; Hassell, 1991; Hogarth & Barth, 1991; Magd, 2003).

Smola and Sutton (2002) found in their study on generational differences in work values that older workers were found to be more loyal to the company and less "me" oriented than their younger counterparts. In order to improve productivity, organizations need to provide a workplace that blends training opportunities, flexible employment patterns, and policies supportive of the life needs of an aging workforce (Stein, Rocco, & Goldenetz, 2000). In order to take advantage of the aging workforce, organizations and HRD practitioners need to be challenged to develop employment policies and practices which will attract older workers.

The hiring and prolonged employment of an appreciable number of older workers in the restaurant industry could go a long way toward dispelling the pervasive stereotype of older workers, to the extent that these older workers prove themselves adaptable, flexible, capable of learning, and technologically current (Rix, 2002). The restaurant industry will be relying on the use of older workers as the available workforce continues to age, and it will be critical to analyze the perceptions of managers regarding this segment of the workforce in order to help influence change in the way that potential employees are recruited and hired.

Purpose of Study

The proposed research has been developed in order to respond to the following research question:

1. What are the perceptions of managers regarding the advantages and disadvantages of hiring older employees in the U.S. restaurant industry?

This question can determine the perceptions of managers regarding the potential of quick service restaurants to hire older workers and the implications that this could have on the organizational development policies and practices regarding recruiting and retention in this steadily growing and labor intensive industry.

METHODOLOGY

The current study was exploratory in nature and gathered and analyzed data from a convenience sample of 55 small to medium sized quick service restaurant organizations in large metropolitan areas in the southeastern U.S. The survey was administered to the management of small to medium sized quick service restaurant chains and franchise groups of quick service restaurant chains. The sample was used in order to determine the perceptions and attitudes of the top management in the organization related to the hiring and employment of older workers in the organization. Simerson (1985) recommends a minimum of 20 supervisors be utilized to develop a pattern for a given occupation.

The survey was designed and adapted from the Magd (2003) study done in Scotland with small to medium hospitality organizations. The survey asked managers for their perceptions regarding specific characteristics of older workers using a Likert scale of 1 = strongly disagree to 5 = strongly agree. The items for the survey contained statements such as: older workers have fewer accidents on the job; older workers are harder to train on the job; I feel that older workers are more dependable. The survey items were qualitatively reviewed by 10 managers in the quick service restaurant industry prior to being administered to the sample to ensure their relevance in the industry.

The data was analyzed using SPSS software (version 13.0) and suggestions were made regarding the findings of the current study relative to the perceptions of managers in these organizations.

RESULTS

The surveys were sent out to 55 restaurant general managers and multi-unit managers of quick service restaurant organizations in metropolitan areas of the southeastern U.S. Of the 55 surveys administered, 21 were returned with one having incomplete data, resulting in 20 useable surveys or a response rate of 36.36%. The results of the study show that there are many preconceived perceptions regarding the hiring and employment of older employees held by management of quick service restaurant organizations.

The characteristics of the respondent organizations are as follows: the organizations ranged in size from one unit to 31 units, with the average organization comprised of 5.40 units. The number of hourly employees in the organizations ranged from a minimum of six to a maximum of 600, with a mean of 139.8 hourly employees. The organizations ranged in the number of managers employed from a minimum of 0 to a maximum of 90, with a mean of 18.5 managers. The number of hourly employees over the age of 55 ranged from a minimum of 0 to a maximum of 60 with a mean of 10.5, and the number of managers over the age of 55 ranged from 0 to a maximum of 5 with a mean of .6 (see Table 1).

The managers were asked a question regarding their expectation of the number of older workers they would be employing in the future. Seventy-five percent of managers surveyed believed that they were going to be employing the same percentage of older employees in the future as they currently employed. Only 10% of the respondents believed that they were going to be employing a larger percentage of older workers in the future than what they were employing now. There were also 15 percent of the respondents that believed that they were going to be employing a smaller percentage of their workforce with older employees in the future than what they currently employed.

The survey was comprised of a total of 23 items. The first four questions related to the employment characteristics of the organization itself and the other 19 items related to the managers' perceptions of older workers relative to specified characteristics. Managers rated the survey items on a Likert-type scale where 1 = Strongly disagree, 2 = Disagree, 3 = Neutral, 4 = Agree and 5 = Strongly Agree. These items rated the perceptions held by the managers regarding older employees in their quick service restaurants. Some examples of these

TABLE 1. Characteristics of Respondent Organizations

	Minimum	Maximum	Mean	Standard Deviation
Number of Units	1	31	5.4	8.27
Number of Managers	0	90	18.5	26.22
Number of Hourly Employees	6	600	139.8	191.87
Managers over 55	0	5	.6	17.92
Hourly Employees over 55	0	60	10.5	1.39

$n = 20$

items were: older employees have fewer accidents on the job; older workers are harder to train on the job; older workers are absent more often than younger employees. The results of these item ratings are shown in Table 2.

TABLE 2. Frequency Distribution of Management Perceptions of Older Workers

Characteristics of Older Workers	Agree and Strongly Agree %	Neutral %	Strongly Disagree and Disagree %	Mean	Standard Deviation
Self motivated, disciplined and respect	80.0	15.0	5.0	4.00	.795
Good communication skills and credibility	75.0	25.0	0.0	3.90	.641
More dependable	75.0	15.0	10.0	3.85	.875
More loyal	70.0	20.0	10.0	3.80	.894
Positive image	75.0	25.0	0.0	3.80	.523
Equal ability to learn	65.0	20.0	15.0	3.70	.979
Have higher quality	60.0	25.0	15.0	3.65	1.137
Cooperate more	57.9	36.8	5.3	3.63	.955
Happier/more satisfied on the job	55.0	45.0	0.0	3.60	.598
Better employees	30.0	65.0	5.0	3.30	.657
More serious accidents	35.0	45.0	20.0	3.15	.933
Hard time adopting new technology	35.0	40.0	25.0	3.15	1.309
Fewer accidents	30.0	55.0	15.0	3.05	.887
Prefer older worker	25.0	50.0	25.0	2.95	.999
Can't keep up with the pace	25.0	40.0	35.0	2.80	.951
High costs	15.0	45.0	40.0	2.80	1.005
Inflexible and reluctant to change	15.8	47.4	36.9	2.79	.918
Harder to train	20.0	25.0	55.0	2.50	1.000
Absent more than younger workers	0.0	5.0	95.0	1.70	.571

$n = 20$
1 = Strongly disagree, 2 = Disagree, 3 = Neutral, 4 = Agree, and 5 = Strongly Agree

When comparing the results of the current study to the Magd (2003) study done in Scotland with 21 managing directors of small and medium hospitality organizations rather than specifically with quick service restaurant organizations, there are differences in the perceptions of the two sets of managers related to the following characteristics: older workers have fewer accidents, preference for older workers, older workers are happier/more satisfied on the job, older workers make better employees, and there are higher costs associated with the employment of older workers (see Table 3). These differences in the results may be a direct result of the cultural differences due to the locations of the studies (Scotland and the U.S.) and also due to the differences in the hospitality organizations surveyed (various types of hospitality organizations and quick service restaurants).

The Magd (2003) study found that managers agreed and strongly agreed to a much larger extent (80.8%) that older workers were happier and more satisfied on the job than the current study (55.0%). The Magd (2003) study also found that the managers perceived that older workers had fewer accidents and were better employees than the current study found. The current study agreed or strongly agreed that older workers had higher costs associated with them at a fairly low level (15.0%) when compared with the previous Magd (2003) study (66.6%). Overall the Magd (2003) study found that the managers agreed or strongly agreed at a much higher level (76.2%) that they would prefer older workers in their organizations when compared to the current study (25.0%).

TABLE 3. Comparison of Current Study and Magd (2003)

Characteristic	Agree and Strongly/ Agree %	Magd (2003)	Neutral %	Magd (2003)	Strongly Disagree and Disagree %	Magd (2003)
Happier/More Satisfied on the Job	55.0	80.8	45.0	4.8	0.0	14.4
Better Employees	30.0	80.6	45.0	19.2	20.0	4.5
Fewer Accidents	30.0	90.4	55.0	0.0	15.0	9.6
Prefer older worker	25.0	76.2	50.0	4.6	25.0	19.2
High costs	15.0	66.6	45.0	14.2	40.0	19.2

Current study ($n = 20$)
Magd (2003) study ($n = 21$)

DISCUSSION

The current study shows that overall management perception is favorable regarding characteristics of older workers. The respondents perceived that older workers were self-motivated, disciplined and showed respect for authority. The respondents also agreed or strongly agreed (75%) that older workers had good communication skills and credibility with customers in the quick service restaurant industry. They believed that older workers were more dependable and that they were absent less frequently than younger workers in quick service restaurant jobs.

Despite the fact that the respondents tended to believe that older workers were more loyal, dependable and that they had equal ability to learn the job as younger workers had, only 25% of respondents stated that they would prefer to have older workers working for them in the quick service restaurants. Fifty percent of the respondents stated that they were neutral when asked if they preferred employing older workers in their organizations. This result implies that despite the fact that older workers possess characteristics that are viewed positively for management, there are perceived barriers in the organizational culture, policies and procedures when it comes to actually hiring older workers.

The majority of respondents believe that they will be hiring older workers in the same numbers as they currently do. With the aging demographics in the U.S., the quick service restaurants in the future will need to ensure that they are looking for ways to change perceptions and encourage the employment of greater numbers of older workers in the industry.

The study shows that there is a perception by managers that older workers possess many positive characteristics. The characteristics of older workers that were viewed as positive by the respondents were that older workers are: self-motivated, disciplined and have respect for authority; they have good communication skills and credibility with customers; they are dependable, loyal and create a positive image for the restaurant; they have equal ability to learn skills and tasks, and their work is of higher quality.

IMPLICATIONS FOR PRACTITIONERS

The current study explores the perceptions of the people in quick service restaurant organizations that make policy and set the tone for the culture of the organizations that they own and/or operate. The perceptions regarding older workers show that there are not many employment characteristics that are perceived to be negative or that are done more thoroughly by younger employees. It is believed by the respondents in the current study that older workers are more loyal and dependable, they have an equal ability to learn in the industry and they have credibility with customers. Why then don't organizations act on these perceptions and actively recruit and attract older workers to their organizations?

The quick service industry is a whole will continue to experience sales increases and increases in the demand for labor and hourly employees (National Restaurant Association, 2005). There will also be an increase in the number of older workers looking for part-time and full-time jobs to supplement the social security or retirement funds that they have accumulated that are not stretching to fill all of the financial needs that they may have (Prewitt, 2005). It seems to be a natural fit between the need for labor and the need for jobs, but there is a concern that the quick service restaurant industry will continue to search for staff that are younger, have good health and "the body beautiful" in order to staff the restaurants (Prewitt, 2005, p. 123).

The restaurant industry has not incorporated their positive perceptions into positive actions quite yet regarding actively recruiting older workers and looking at how the work is done in order to make it positive for both the organization and the older workers. The implications from the current study are that managers need to look at their current practices regarding recruiting and hiring older workers and work to change the culture in their organizations in order to make it more accepting of older employees. There is a disconnect between the perceptions of managers regarding older employees and the actual HRD practices regarding actively trying to hire older workers. Organizational development plans in quick service restaurant organizations need to include changing hiring practices in order to match the management perceptions regarding older workers. With the struggle of the quick service restaurant industry in finding employees, the managers need to work toward changing paradigms of the past.

LIMITATIONS OF CURRENT STUDY

The current study is not without limitations and requires future studies in order to ensure that the results can be generalized to other organizations. One of the biggest limitations of the current study is the small sample size of the number of quick service restaurant managers and organizations. It is very limiting due to the small geographical area represented as well.

Another limitation is that it is asking managers about their perceptions regarding older workers and the hiring of older workers, but it is not longitudinal in nature and therefore the actual hiring practices were not observed. The possibility exists that the managers interviewed do not realize the extent of the labor issues and do not know for certain how many older employees will be hired in the future.

CONCLUSION

The current study looks at quick service restaurant organization management perceptions of a select group of small to medium restaurant companies. It presents the findings as well as some implications for practitioners regarding ways that quick service restaurants can take the increase in the number of older workers in the workforce and try to fill the gap left by the decrease in the number of younger workers. The younger worker has been the traditional employment mainstay of the quick service restaurant industry and in the perception of the respondents of the current study, younger workers will continue to be the employment and recruiting focus. The responses of the current study show that there are many perceived benefits of older workers, but the traditional mindset has been to ignore this growing segment of the workforce.

The human resources policies and practices of organizations need to ensure that older workers as a segment of the labor market is not ignored when creating policies for recruiting and staffing. The quick service restaurant industry appears to be a segment that is continuing to try to staff the restaurants the same way that they have since the inception of the industry in the 1950's. In the current study, the average percentage of older employees and older managers working for the organizations is approximately 7.5% and 3.24% of the employment ranks respectively. These numbers do not reflect the current age demographics of the U.S. Organizational development programs need to ensure that there are various ways to support the growth in the industry by looking to a segment of the population that has been previously untapped.

In order for the quick service restaurant industry to be able to continue to work toward full employment in the restaurants in a productive way, and to take advantage of the perceived benefits of older worker characteristics, as shown in the current study, organizations need to be proactive. There needs to be some thought and planning for how older workers can add some dependability and maturity into the restaurants, and can possibly change part of the image of quick service restaurants. Traditionally, quick service restaurants have been staffed with younger employees and have been desperately seeking more employees for years. Working to actively recruit and hire older employees could add a new dimension and some new markets for quick service restaurants.

Future research should be directed at the expanding the current exploratory research to explore a wider cross section of the quick service restaurant industry, as well as to look at the restaurant industry as a whole. Older people are a willing and able part of the labor pool that is expanding and not receiving the credit due them in this segment of the hospitality industry.

REFERENCES

American Association of Retired Persons (AARP) (2001). Beyond fifty: America's future. Retrieved on May 10, 2005 from *http://www.aarp.org/about_aarp/aarp_leadership/on_issues/aging_issues/a2002-12-31-novellicleveland.html.*

American Association of Retired Persons (AARP) (2000). American business and older employees: a summary of findings. Retrieved on September 2, 2005 from http://assets.aarp.org/rgcenter/econ/amer_bus_findings.pdf.

American Association of Retired Persons (AARP) (1998). *Boomers Look Toward Retirement.* Washington, D.C.

Barth, M. C. (2000). An aging workforce in an increasingly global world. In C Morris & E. Norton (Eds.), *Advancing Aging Policy as the 21st Century Begins*. New York: Haworth.

Barth, M. C., & McNaught, W. (1991). The impact of future demographic shifts on the employment of older workers. *Human Resource Management, 30*(1), 420-434.

Bolch, M. (2000). The changing face of the workforce. *Training*, December, 73-78.

Burke, W. W. (1994). *Organization Development: A Process of Learning and Changing* (2nd ed.). Reading, MA: Addison-Wesley.

Burtless, G. (1998). Increasing the eligibility age for Social Security pensions. *Testimony for the Special Committee on Aging.* United States Senate, July 15, 1998.

Caterer and Housekeeper (1992). We can not afford ageism. 11 June, p. 9.

Catrina, L. (1999). Older workers: flexibility, trust and the training relationship. *Education and Ageing, 14*(1), 51-60.

Cheney, S. (2001). *Keeping Competitive: Hiring, Training, and Retaining Qualified Workers.* Washington, D. C.: Center for Workforce Preparation.

Daft, R. L. (1998). *Organization Theory and Design* (6th ed.). Cincinatti, OH: South-Western College Publishing.

DeMicco, F. J. (1989). Attitudes toward the employment of older food-service workers. *Hospitality Education and Research Journal, 13*(3), 15-30.

DeMicco, F. J., & Reid, R. D. (1988). Older workers: A hiring resource for the hospitality industry. *The Cornell Hotel and Restaurant Administration Quarterly*, May, 56-61.

Dibden, J. & Hibbert, A. (1993). Older workers: an overview of recent research. *Employment Gazette*, June, 237-249.

DiPietro, R. B., & Milman, A. (2004). Hourly employee retention factors in the quick service restaurant industry. *International Journal of Hospitaity & Tourism Administration*, *5*(4), 31-51.

Doering, M., Rhodes, S. & Schuster, M. (1983). *The Aging Worker*. Beverly Hills, CA.

Doverspike, D. (2000). Responding to the challenge of a changing workforce: Recruiting nontraditional demographic groups. *Public Personnel Management, 29*(4), 445-458.

Gibson, K., Zerbe, W., & Franken, R. E. (1993). The Influence of rater and rate age on judgements of work-related attributes. *The Journal of Psychology*, *127*(3), 271-281.

Haider, S., & Loughran, D. (2001). "Elder Labor Supply: Work or Play?" Paper presented at the Third Annual Joint Conference of the Retirement Research Consortium, Washington, D. C.

Hassell, B. (1991). A causal model examining the effects of age discrimination on employee psychological reactions and subsequent turnover intentions. *International Journal of Hospitality Management*, *10*(3), 245-258.

Herman, R. E., Olivo, T., & Gioia, J. L. (2003). *Impending Crisis*. Winchester, VA: Oakhill Press.

Hignite, K. (2000). Aging gracefully. *Association Management*, *52*(8).

Hogarth, T. & Barth, M. (1991). Costs and benefits of hiring older workers: a case study of B&Q. *International Journal of Manpower*, *12*(8), 5-17.

Kaeter, M. (1995). Un-retirement. *Training*, *32*(1), 63.

Lyon, P., & Mogendorff, D. (1991). Grey labour for fast food. *International Journal of Hospitality Management*, *10*, 25-34.

Magd, H. (2003). Management attitudes and perceptions of older employees in hospitality management. *International Journal of Contemporary Hospitality Management*, *15*(7), 393-401.

McCool, A. C. (1988). Older workers: Understanding, reaching, and using this important labor resource effectively in the hospitality industry. *Hospitality Education and Research Journal Proceedings*, *12*(2), 365-378.

Meyer, R., & Meyer, G. (1988). Older workers: Are they a viable labour force for the hotel community? *Hospitality Research Journal*, *12*(2), 361-364.

National Adult Tracking Surveys. (2001). *Referenced in Burger King Corporation meeting, February, 2002*, Harriet Gallu.

National Restaurant Association. (2002). *Quickservice Restaurant Trends 2002*. Washington, D. C.: National Restaurant Association.

National Restaurant Association (2003). State of the Restaurant Industry Workforce: An Overview. Retrieved on May 15, 2005 from *http://www.nraef.org/solutions/pdf/RestaurantIndustryWorkforce.pdf*.

National Restaurant Association (2005). Restaurant Industry 2005 Fact Sheet. Retrieved on May 5, 2005 from *http://www.restaurant.or/pdfs/research/2005factsheet.pdf*.

Perry, P. (2005, May). Getting gray to stay. *Restaurant Hospitality*, May, 2005, 101-104.

Prewitt, M. (2005). Not the retiring kind: Aging workforce risks bias despite industry's job needs. *Nation's Restaurant News*, *39*(15), pp. 1, 123.

Rix, S. E. (2002). The Labor Market for Older Workers. *Generations* (Summer), 25-30.

Rocco, T. S., Stein, D., & Lee, C. (2003). An exploratory examination of the literature on age and HRD policy development. *Human Resource Development Review*, *2*(2), 155-180.

Sargeant, M. (2001). Lifelong learning and age discrimination in employment. *Education and the Law*, *13*(2), 141-155.

Schwartz, P. (2003). *Inevitable Surprises*. New York, NY. Penguin Group.

Simerson, G. (1985). Minnesota Job Description Questionnaire. In Keyser, D. & Sweetland, R. (Eds.), *Test Critiques: Vol 2*. Kansas City, MO: Test Corporation of America.

Smola, K. W., & Sutton, C. D. (2002). Generational differences: Revisiting generational work values for the new millennium. *Journal of Organizational Behavior*, *23*, 363-382.

Speilberg, S. (2004). Forecast for year: Registers to ring, economy in swing. *Nation's Restaurant News*, *38*(1), 29-30.

Stein, D. & Rocco, T. S. (2001). The older worker. Myths and realities (Report No. 18). Columbus, OH: ERIC Clearinghouse on Adult, Career, and Vocational Education. (ERIC Document Reproduction Service No. ED459361).

Stein, D., Rocco, T. S., & Goldenetz, K. A. (2000). Age and the university workplace: A case study of remaining, retiring, or returning workers. *Human Resource Development Quarterly*, *11*(1), 61-80.

Strate, M. & Torraco, R. (2005). "Career Development and Older Workers: A Study Evaluating Adaptability in Older Workers Using Hall's Model." Paper presented at the Annual Conference of the Academy of Human Resource Development, Estes Park, CO.

Tayler, P. & Walker, C. (1993). Employers and older workers. *Employment Gazette*, August, 371-378.

Thatcher, M. (1996). Bouncing back from the scrapheap at 41. *People Management*, 11 January, p. 19.

Tomb, D. (1984). *Growing Old*. Viking Press, New York.

U.S. Department of Labor, Women's Bureau (DOL) 2005. Quick facts on older workers. Retrieved from *http://www.dol.gov/wb/factsheets/Qf-olderworkers.htm*.

Wade, K. W. (2003). *Marketing the Workplace to Older Workers: Determinates of Employer Choice Among Mature Employees*. Argosy University, Sarasota, FL.

Yeates, D., Folts, W., & Knapp, J. (2000). Older worker's adaptation to a changing workplace: Employment issues for the 21st century. *Educational Gerontology*, *26*(6), 565-583.

doi:10.1300/J369v09n02_10

Comparison of Internship Experiences in Food Service Firms in India and UK

Vinnie Jauhari
Kamal Manaktola

SUMMARY. People are the most important asset for a service firm. Internships could be used as a good source for identifying potential talent in hospitality firms. Internship experiences could be used for creating memorable experiences as an employee and as a potential customer. The study assesses internship experiences of interns in UK and India. Internship experiences have been compared on content of the training, mentorship, hygiene factors, rewards and recognition. A model for managing internship experiences for interns in hospitality industry is suggested. The paper has policy implications for training and recruitment directors in hospitality industry and academic institutions. doi:10.1300/J369v09n02_11 *[Article copies available for a fee from The Haworth Document Delivery Service: 1-800-HAWORTH. E-mail address: <docdelivery@haworthpress.com> Website: <http://www.HaworthPress.com> ©2006 by The Haworth Press, Inc. All rights reserved.]*

Vinnie Jauhari, PhD, is Professor and Head, School of Management & Entrepreneurship, Institute for International Management & Technology, # 336, Udyog Vihar, Phase IV, Gurgaon-122001 India (E-mail: vjauhari@iimtobu.ac.in).

Kamal Manaktola, MA, is Head, School of Hospitality & Tourism Management, Institute for International Management & Technology, # 336, Udyog Vihar, Phase IV, Gurgaon-122001 India (E-mail: kamal@iimtobu.ac.in).

[Haworth co-indexing entry note]: "Comparison of Internship Experiences in Food Service Firms in India and UK." Jauhari, Vinnie, and Kamal Manaktola. Co-published simultaneously in *Journal of Foodservice Business Research* (The Haworth Hospitality & Tourism Press, an imprint of The Haworth Press, Inc.) Vol. 9, No. 2/3, 2006, pp. 187-206; and: *Human Resources in the Foodservice Industry: Organizational Behavior Management Approaches* (ed: Dennis Reynolds, and Karthik Namasivayam) The Haworth Hospitality & Tourism Press, an imprint of The Haworth Press, 2006, pp. 187-206. Single or multiple copies of this article are available for a fee from The Haworth Document Delivery Service [1-800-HAWORTH, 9:00 a.m. - 5:00 p.m. (EST). E-mail address: docdelivery@haworthpress.com].

Available online at http://jfbr.haworthpress.com

doi:10.1300/J369v09n02_11

KEYWORDS. Internship experiences in India, UK, empowerment, stakeholders, customers, job responsibility

INTRODUCTION

A sustained level of growth in an organization requires concern for all stakeholders. Employees are important stakeholders who help in building up an organization's image. Infact, employee satisfaction can be an important component contributing towards profitability and enhanced market share. Employee satisfaction ultimately leads to customer satisfaction (Jauhari, 2001). Internships can be the first point of contact between the firm and a prospective employee. Hence internship experiences are important and therefore need attention. There is a moment of truth (Zeithaml & Bitner, 2000) for a consumer which plays an important role in shaping the entire service experience. Similarly, there is a moment of truth for an intern and images built in during first contact will influence an employee's attitude towards a firm and eventually as a consumer as well. Firms spend millions of dollars on PR. Managing image and creating favourable experiences for interns can play an immensely important role in shaping a firm's image. A structured internship can also be indicative of good HR practices and communicate about a firm's commitment towards legal and ethical practices. This study assumes significance as it attempts at understanding internship experiences across Indian and British cultures and develops a model for enhanced internship experience for interns.

OBJECTIVES

The study for documenting internship experiences has been conducted with the following objectives:

- Compare internship experiences of undergraduate hospitality students in foodservice firms across Indian and British cultures.
- To evolve a framework for creating positive internship experiences both for interns and employers.

METHODOLOGY

The study has been carried out in India and UK using both secondary and primary research. The study deployed a questionnaire based survey for interns who were pursuing undergraduate degree program in hospitality. The sample comprised of two groups. One group of interns were pursuing internship program from a private hospitality institute in India. The other group of interns were pursuing the hospitality program from a well known British University. As a part of the undergraduate degree from this British University, it is mandatory for interns to do a 48 week training. Convenience sampling was used to identify interns who had undergone internship in India and UK. A group of 80 interns (40 having done internship in India and 40 having done internship in UK) were identified. A usable set of 57 questionnaires were used for the study. A group of 30 interns who have undergone internship in India and another group of 27 interns who have undergone training in UK constitute the final sample. Structured questionnaires were used to assess the internship experiences of these students.

The questionnaire was prepared which explored different dimensions of internship such as content of assignments, compensation, nature of mentorship, structure of training and explored possibilities for future association as customers and employees. The questionnaires were tested for content validity by three experts in the area of HR in hospitality industry. Pre-testing of the questionnaire was done with a group of randomly selected students who had undergone training in either India or UK. The questionnaire based study was also supported with interviews conducted with ten hospitality experts and similar number of interns about internship experiences. This helped in finding out qualitative information about internship experiences across Indian and UK cultures. Data so collected was analyzed both qualitatively and quantitatively. A conceptual model was proposed using the qualitative analysis both from questionnaire and interview based survey. Factor analysis was used to understand internship experiences of interns in hospitality industry. The findings so generated help to make recommendations on enhancing internship experience for interns. These could be implemented both by the Hospitality firms as well as educational institutes managing hospitality programs.

REVIEW OF LITERATURE

Internationally, tourism industry is poised for growth. This has implication for growth in the hotel and foodservice establishments as well. There will be a surge of demand in employees for foodservice establishments. In US, according to National Restaurant Association, there are 12 million jobs in lodging and food service, and eight million more jobs in businesses that supply the industry (Carmen, 2002). The tourism sector in India has a potential to create employment opportunities for 25 million people according to WTTC statistics (Jauhari, 1999). In this scenario, it is therefore pertinent that we look at the cost element for recruitments. Interns can be a low cost option for recruitment and also a means of containing costs as well if well structured plans are in place for managing internships. Internships are of growing value to faculty and hospitality organizations and the continued development of experiences must be a priority (Harris and Zhao, 2004). In US, employers love internship programs as they function as pre-employment programs. Internships also help to cut hiring costs which typically range from $2,500 to $5,000 for advertisements, interviews and training (HR magazine, 1997). Internships are of growing value to faculty and hospitality organizations and the continued development of experiences must be a priority (Harris and Zhao, 2004). An interesting suggestion made by hosting executives was the hope that they could arrange internships with educational institutions. However the training experiences are very varied. They range from actual job assignments to ordinary assignments. For instance, foreign interns hired through American Hospitality Academy go through different experiences. Mariano (2002) label these interns as well trained, polite, have positive attitudes and hope to impress. They nurture a dream of becoming leaders in the hospitality industry.

Downey and DeVeau (1988) have elaborated on hospitality internships and recruiters expectations from hospitality educators.

Internships Advantages to an Employer

Internships offer numerous advantages to an employer. They help to cut costs as internship wages are lower than wages offered to full time employees. Interns have lot of enthusiasm to work. They also would not have issues of running away from their responsibilities. Internships also help to find future employees. Having the right fit between organization culture and attitude of an employee is the biggest challenge. Internship provides an opportunity to an employer to understand the skill and knowledge base of an individual. It is an investment to hire a future employee without having incurred the risk of hiring a wrong person. Internship experiences could also provide opportunities to know a future employee. An organization can use this as a subtle experience for a future employee and a customer. Spreading positive word of mouth through the own employees can create a huge brand equity. Interns can be a valuable resource to get fresh insights into the processes of a firm. People who are engaged themselves in a process sometimes fail to see flaws which an outsider can see. An idea about internal processes can bring back a small band of loyal customers. Firms talk of loyalty programs but there is also a captive chunk which firms can exploit. Interns must also be seen as future customers who would create brand equity in their circles.

Internship Advantages to Students

Internships offer numerous advantages to interns such as exposure to work place. The internship offers opportunity to understand the work place better. Future managers get an opportunity to understand processes and systems. They are able to see linkages between various departments and understand value systems as well. Internship opportunity offers an opportunity to link the practice with theory. Learning is immense at the workplace as there are always differences between reality and what is taught. Insights from customers perspectives helps them see things differently. Internship experiences exposes the interns to reality of the industry. It also helps them to understand their own selves. Some of the internship experiences put off interns completely and they subsequently think of higher education and shifting away from the industry. In India some these shifts have been witnessed. Internship experience also brings in clarity on areas of interest as well. Some one may prefer a career in kitchen while as some feel they are more cut out for HR or finance related jobs. Any amount of classroom teaching cannot be a substitute for real life experience. Interns by way of real life experience get exposed to factors which yield better returns for the business firms. Also an exposure for reasons for failure can also act as a seedbed for future entrepreneurship. Some students while as interns conceive of new business ideas resulting in new ventures. This also helps them to form the right values and creates an understanding the essentials to move up in the organization. It opens doors for future employment opportunities for interns. If they have demonstrated commitment and the right values, it offers them future position with a firm that they have worked for.

Cross cultural internships are even a higher learning experience on account of the following reasons: tremendous learning about a new culture and work practices; dealing with diverse work force issues; dealing with guests from a different culture; understanding differences between service standards in the home and the foreign country; expectations in international workplace (Tas, 1988; Okeuji et al., 1994); benchmark performance; coping mindset and develops survival instincts as well. Internships that are a valuable experience to the student involve considerable investment of time. Fully accredited schools generally insist on some sort of up-front written agreement between a company and the student as to what is expected of both parties before agreeing to give credit for internships (Strock, 1991). For internship to be successful, both employer and employee must share the same perceptions about the internship and what the student can bring to internship experience. Experience suggests that employers' and interns' perceptions tend to vary along a number of dimensions. These include conflicting perceptions of what constitutes ethical behaviour, the intern's written and oral communication skills, what constitutes acceptable office behaviour and the intern's technical skills (Trackett, Wolf & Law, 2001).

FINDINGS

Internship programs in India typically have a shorter duration of about six months while as students in UK go for almost an year's training. The disparities come in on account of differing curriculum in academic programs as well. The training in UK is more structured with students actually assuming specific job roles such as events manager, HR manager, F&B assistant manager while as in India, these assignments are really unheard of in food service organizations. In India and UK, most students get exposed to more than two departments during internship. This gives them a good exposure and an idea about work systems across various departments. The inputs on various aspects of internship are documented in Table 1.

Internship programs are more structured in UK as compared to India. There is a disparity in compensation provided to interns. The local laws in UK ensure a minimum wage rate. However, in India the compliance is not there. Almost 56% of interns in the sample did not get any wages during internship. There is also a huge disparity in the wages offered to students during internship. In UK, the minimum wage rate ensures that students earn enough to survive and take care of their local expenses. In India, experience are very diverse. An interview with some key hospitality professionals reveals that interns are seen as a liability by most firms. In fact in some of the leading public sector firms, interns had to pay a particular amount to ensure that they had an internship. Lack of adequate wages leads to lack of seriousness on the part of the intern as well as employer. Indian experience indicates lack of pre-designated work assignments for interns. As contrast to this internships in UK hotels offer better opportunities in terms of job content and empowerment. Students

TABLE 1. Induction Experiences of Interns in India and UK

	India		UK	
	Yes	No	Yes	No
1. Compensation	17	13	27	0
2. Overtime requirement	25	5	27	0
3. Payment for overtime	2	28	27	9
4. Accommodation Support during Internship	2	28	24	3
5. Learning Support from Supervisors	22	8	26	1
6. Discussion on Professional issues with supervisors	23	7	23	4
7. Value awareness about the workplace	14	16	23	4
8. Future Customers	24	6	21	6
9. Work as a future employee	19	11	23	4
10. Rewards	14	16	19	8
11. Independent work assignment	20	10	27	0
12. Odd jobs	18	12	9	19
13. Happy with work culture	19	11	24	3
14. Offer to work After Internship	15	15	25	2

actually independently handle real time assignments. This puts in a pressure for performance and also students tend to perform with sincerity. The lack of appropriate wages in Indian context is also on account of lack of delegation of proper assignment to hospitality interns in India. They are seen as a liability rather than as a resource who could be a potential employee or a future customer. Interns in UK make money on overtime as well which gives them an opportunity for cross functional training as well. The payment for overtime is not made in India in 93% cases while as in UK it is mandated by law. The overtime option is exercised by foreign students as it helps them to contribute to their earning and gives them more work exposure. The overtime wage differentials are there in UK as well depending upon the hotel property in which a student works. The normal working hours for most students in India in hospitality firms is more than 8 hours. However in UK, the prescribed hours mandated by law are followed.

In UK, accommodation support is provided in most of the internship assignments while as in India, interns are expected to make their own arrangements. This coupled with low wage rates along with lack of overtime wages makes it extremely difficult for students to go through internship. It is viewed as a hardship and exposes students to inhospitable work environment. Internship experiences in both India and UK indicate that there is a learning support from the supervisor. However, in case of UK an actual job assignment makes it a worthwhile experience for an intern.

There is a distinct difference in awareness levels of work values in India and UK (Refer to Table 2). Contrary to internship experience in UK, wherein interns were aware of the workplace, in India this awareness is restricted to less than 50%. This implies that the employees and interns are not aligned to the demands of the business. The analysis of values listed by interns in India and UK give some important insights into work ethics and work environment in hospitality firms. In UK, guest satisfaction, employee satisfaction and team work are considered important values while as in India aspects such as hard work, sincerity and honesty are shared more commonly. This also implies that operationalisation of the values in management terms may or may not be there.

Eighty percent of Indian students would want to visit the firm as a customer as against 77% in case of US (Refer to Table 1). The data indicates that 85% of interns in the sample in UK would like to come back to work as an employee as against 63% in case of Indian hospitality firm (Table 1). This reflects that 37% of interns would not like to come back to the same work place as employees. This implies that the Indian hospitality firms need to re-look at the way they treat interns. It is pertinent to understand the aspects of workplace which contribute to a positive work experience.

TABLE 2. Work Values in UK and India

Values considered of importance in UK	Values considered of importance in India
100% guest satisfaction	Hard work
Employee satisfaction	Sincerity
Team work	Honesty
Positive cash flow	We know what it takes
Can do attitude	Moment of truth
100% service guarantee	Personal touch
Customer care	The guest is always correct
Profit maximization	Discipline
I can do attitude	Job satisfaction
Staff friendliness	Welcome guest with a smile
	Increasing employee morale
	Services is our motto
	Guest satisfaction

Table 3 highlights positive dimensions of the internship experiences in India and UK. The Table 3 implies that in Indian context expressions such as, work environment, staff, work culture are used while as interns having experienced an internship in UK tend to look at dimensions such as supporting staff, learning and knowledge, good working conditions. So they are higher on learning curve both in terms of skill development and exposure. In UK, students liked supportive staff, global exposure, learning and knowledge, good atmosphere related with the work culture. The Table 1 indicates that only 66% cases in India, interns are entrusted with independent work assignment as against 100% in UK. In UK, just 30% interns admitted that they did odd jobs a well during internship as against 60% in case of India. Sixty-three percent of students in India were happy with the work culture as against 88% in UK.

Table 1 indicates that 47% interns in India are rewarded for outstanding performance as against 70% in case of UK. The mode of rewards are quite different in Indian and UK system for the interns. The various modes of rewarding are indicated in Table 4.

TABLE 3. Aspects of Work Culture in UK and India That Students Liked

Aspects (In descending order) for UK	**Aspects (In descending order) for India**
Supporting staff	Work environment
International exposure	Staff
Learning and knowledge	Industrial exposure
Good atmosphere	Work Culture
Industry experience	Work under pressure
Good working conditions	Casual supervisor
Team responsibility	Industrial exposure
Confidence	Local competition
Customer satisfaction	Clear picture of working in a hotel in India
Colleague satisfaction	Intensive product knowledge
Excellent team	Local competition
Know more about a different culture	Brand name
Brand name	Better employee coordination
Training support	
Working patterns	
Pressure handling	
Hotel policies	
Equal opportunity	
Time management	

TABLE 4. Mode of Rewards for Interns in UK and India

UK	India
Employee of the month	Rs. 250 as cash incentive
Fish forms	Pat on the back
Star points	Appreciation letter
Certificates	Appraisal letter
Praise by team members	Tips
More authority and control	
Appreciation in public	
Remuneration	
Club membership	
Recognition in Appraisal	
Took training session for others	
Respect	

Table 4 indicates that systems of rewarding employees are more structured in case of UK as compared to India. The various schemes such as employee of the month, fish form, star points are not offered to Indian interns in India.

The changes recommended by UK interns are mainly in areas of logistics. There is also expectation of higher wage rate as well. The following changes are suggested: Food for staff; accommodation support; higher pay scales; better overtime wages. The changes suggested for Indian hospitality internship experience are related with offering stipend during training, food and formalising the training as well. There is also an indication of need for higher degree of empowerment and a better management of work load as well.

Table 5 indicates that customers are happy with the service standards in both India and UK. It demonstrates that 73% of the sample feels that employees were not happy with their compensation. In UK, 66% were happy. This gives an insight into lack of basic benefits in Indian context. This may put off students for subsequent career in hospitality in India. Sixty-two percent of interns in UK feel employees feedback in considered important in initiating change. Also a similar percentage feels that employees are involved. In both cases supervisors tend to motivate interns. The differentials in both cases seem to be negligible. In India, just about 50% are offered jobs as against 92% in case of interns in UK. This implies that firms are looking at internship as a means to recruit and identify potential while as in India the efforts are not channelised well (Table 1). Eighty-three percent of interns in India feel that there is lack of empowerment as against 7% in case of UK trained interns. This

TABLE 5. Intern Experience on Work Culture in India and UK

		Response					
Sl. No.	**Variable**	**Country**	**Never**	**Sometime**	**Neutral**	**Often**	**Always**
1.	Supervisor criticism	India	4	12	7	3	4
		UK	4	16	3	2	2
2.	Consumers happiness with service standards	India	1	2	7	17	3
		UK			5	21	3
3.	Employees happiness with compensation	India	16	6	8	0	0
		UK	3	5	6	11	1
4.	Employee feedback in making changes	India	2	6	6	8	5
		UK	3	9	5	8	4
5.	Supervisor motivation	India	1	6	5	10	7
		UK	0	0	3	11	13
6.	Consistency in systems	India	3	6	7	11	3
		UK	0	1	3	17	5
7.	Empowerment	India	22	3	3	2	0
		UK	0	2	1	14	10

certainly has implications for bringing about change in Indian hospitality industry.

Factor Analysis for Internship Experiences

A factor analysis was carried out using variables on which responses were collected on a semantic differential scale. The factor analysis was carried out using Stata 7.0. The analysis was done using principal component method. The analysis is indicated in the Table 6.

A total of 14 variables were used for 57 respondents. The factors which seem to emerge on account of factor analysis are indicated in the Table 7.

The factor analysis reveals that the factors that explain the internship experience for hospitality can be grouped into four categories:

Factor 1: Work Exposure Duration

This factor includes two variables–duration of training and duration of working hours. Duration of training has a negative relation with training experience. A student under going shorter training assignment is not

TABLE 6. Factor Analysis Output

Factor	Variance	Difference	Proportion	Cumulative
1	1.70	–0.22	0.2583	0.2483
2	1.92	0.33	0.2810	0.5293
3	1.59	–0.38	0.2325	0.7618
4	1.63		0.2382	1.0000

TABLE 7. Factor Analysis for Internship Experience

Variable	Factor Loading				
	1	2	3	4	Uniqueness
Training Duration	–0.48				0.50
Normal working hours	0.97				
Customer happy with standards		0.45			0.75
Employee feedback for making change		0.92			
Happy with work culture			0.49		0.48
Employee happy with compensation			0.94		0.01
No. of departments				0.71	0.46
Empowerment				0.50	0.45

able to learn enough and exposure is limited as well. The working hours also contribute to the whole experience. In case of UK, longer working hours are compensated by overtime payments. However, in India a lack of overtime payment and low compensation contributes negatively to the whole experience. The implication for hospitality HR departments in India is that compliance with normal working hours needs to be there. A stretch beyond normal hours leads to negative perception about the hospitality industry.

Factor 2: Stakeholder Factors

The customer satisfaction and involvement of employees through their feedback for initiating change in the organization are important aspects contributing to a positive training experience. Customer satisfaction creates business growth and also motivates employees for a better performance. Employee involvement boosts self esteem and also generates positive motivation for a superior performance. So hospitality firms need to create positive experiences for firms. This opportunity needs to be utilized fully by the firms to identify and attract outstanding talent. Many other industries such as software firms such as Microsoft, Infosys have fostered close linkages with academic institutions to identify young talent. Indian firms already have challenges for manpower. A well structured internship program could channelise the young talent much more effectively.

Factor 3: Hygiene Factors

These include aspects related with happiness with work culture. It also includes satisfaction levels of employees with compensation. This is in consonance with Herzbeorg's Theory (1959). The implication is that hospitality firms need not only look at the job content but also at the wage levels. In India, inappropriate wages create unfavourable disposition towards hospitality industry. This in subsequent years will manifest as a serious problem. An intern gets demotivated by lack of opportunities for growth when he envisions lack of suitable compensation.

Factor 4: Empowerment Factor

This factor includes exposure in various departments and the power of making independent decisions. Empowerment factor brings out a sense of involvement of employee with the organization. This may create positive images which may bring out potential association of the employee with the firm as a future customer or as an employee.

CONCEPTUAL MODEL FOR MANAGING INTERNSHIPS AT WORKPLACE

The outcome of analysis conducted at the workplace has resulted in development of a conceptual model for managing internships at the workplace. The model that seems to emerge out of this study is that interns should not be looked at as liability by a firm but as an opportunity to create positive experiences for them. A firm expends a lot of money to create positive impressions on customers. Herein a model is suggested for creating positive experiences for interns.

Experiences Prior to Training

The interns can be viewed as future customers, future employees, future vendor, strategic alliance partner or stakeholder or an employee with a stakeholder. All the above stakeholders contribute to building an image for the firm. There is a need to create positive experience for the interns prior to training, during training, after training. Experiences prior to training would include interventions such as–Clarity on internship outcomes, recruitment practices, possible opportunities, setting the right expectations among interns on outcomes of internship. Setting the right expectations are important as creating illusions about workplace leads to higher levels of dissatisfaction eventually and therefore negative experiences.

Experiences During Training

The experiences during training would involve the following factors as enumerated in the Figure 1. These would deal with adopting standard

FIGURE 1. Model of Creating Positive Internship Experiences for Interns

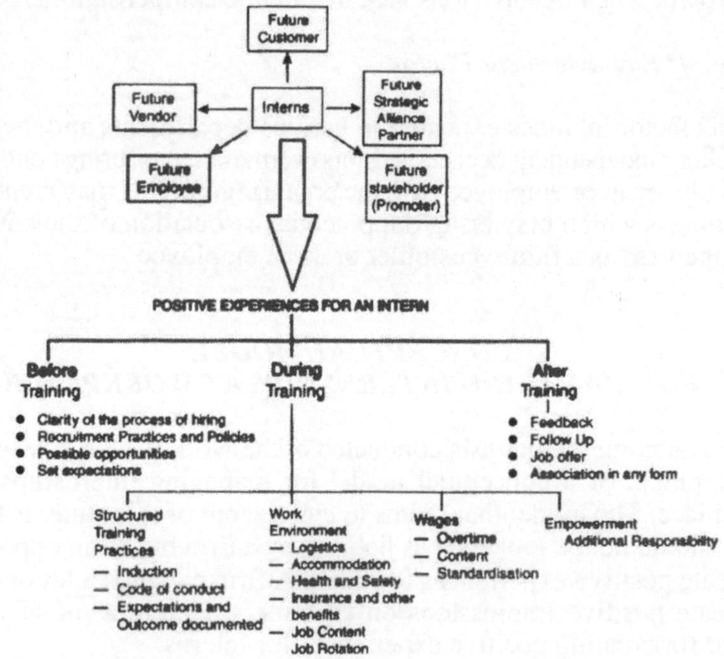

training practices. This would involve induction as well as specification of a code of conduct. This would also include creating the right work environment. A support on logistics and accommodation is helpful specially if an intern works outside the city of residence. The third aspect is following the minimum wage time and having overtime payment norms in place. The content of the training needs to be standardized across various departments. There is also a need for exposing interns to different departments as well as empowering them with decision making powers. All these contribute towards creating positive experience towards interns.

Experience After the Training

After the training is offered, firms could take a feedback to initiate change. There could be a possibility to offer a full time or a past time assignment and channelising interns strengths in some way if desirable.

As an outcome of this model, the following hypotheses are suggested:

- Hypothesis 1: Positive experiences of interns could lead to positive word of mouth for the firm.
- Hypothesis 2: Positive experiences of interns could lead to attracting good employees.
- Hypothesis 3: Positive experience of interns could impact decisions to stay on subsequently in the hospitality industry.
- Hypothesis 4: Positive experiences during internship could provide avenues for unique contribution/innovation from interns.
- Hypothesis 5: Positive experiences could lead to a better understanding of work dynamics from the employers' perspective.

CONCLUSION

Internship experiences of interns could have an impact on their future association with the firm as an employee or as a customer. The study points out that cultural differences play a very important role in creating different internship experiences. While as internship experience in UK indicates higher degree of empowerment and higher degree of involvement of the organization, Indian internship experience indicates that internship is not really structured well. Interns in UK had more independent roles allocated to them at the work place. The compensation was fair and wages for overtime were paid. In Indian context, interns are seen as liability with them with no independent assignment. This attitude puts them off. The compensation is minimal which is not even sufficient to care of their living expenses. Other than this long hours at work and lack of involvement puts them off. Many interns therefore completely lose interest in taking hospitality assignments on completion of the programs. Many interns therefore look for international assignments where they feel there is a higher degree of respect and involvement and wages rates are higher. Indian firms therefore do not make use of opportunity to create positive experiences both as a potential employees and customers. On discussion with Indian hospitality firms, it has been found that the reasons for this kind of a treatment is that the skill set of interns coming in from traditional hospitality schools are not developed enough to put them on jobs involving customer interface. Also within the organizational context, the training program is not well structured. Commitment to managing this well needs to come from the top management. Indian hotels do not take the availability of interns in staff planning resulting in overcrowded organisations when interns are available. Low supervision, lack of induction, no empowerment, internship having limited linkage with intern's learning requirements are evident in this study. In UK, the positions are identified for interns based on the skill set available and the treatment given to the interns is more like an employee who still needs some on-the-job training. Most hotels in India take staff cost as fixed cost and business variation has limited effect on this cost. Interns are treated as extra burden and hence low stipends. In fact, the money paid as stipend is insufficient to even pay for one week's transport expenses. In UK, the interns are not extra staff. Employing interns lowers the fixed element of the staff cost resulting in higher profits. The stipend/wages paid are slightly lower than the regular employees but sufficient for an intern to survive in that country.

It is evident from the study that the students doing their training in India do not express their opinion so freely because of imbalance in supply and demand. All hotel management students in India are required to do internship for a period ranging between six months to one year. Without the internship they are not eligible for the award of the degree/diploma. Since demand for internship in Indian hotels is huge and the requirement is much less than the supply, hotel management institutes often find it difficult to organise internship for their students leaving no choice for the students but to run from pillar to post for arranging their own internship. Under these circumstances, students feel obliged to the organisation that offers them the position. More number of hours, menial jobs, low pay, no proper guidance are directly reflecting towards exploitation of the needy students who have no option but to feel grateful to the organisation providing them the opportunity to get the required exposure (unstructured learning).

Whereas the UK scenario is completely different, UK hotels impart interns to create winning combination, students get international experience and hotels get staff to man the positions. UK hotels are spending their resources in developing these interns for the macro hospitality market and to augment their brand equity in job market.

RECOMMENDATIONS

Most of the recommendations in this paper would be for the Indian hospitality industry and the policy makers in the education sector.

Internship has been provided as a part of the curriculum to develop young student's employability through structured on-the-job training. In Indian hospitality industry this is possible once the links between education and the industry are strengthened. Industry should take the lead to decide on the number of positions that can be reserved for proper structured internship and ensure that interns are provided the required inputs and are groomed to take up hospitality positions once they graduate. Good recruitment practices could help hotels to find the right people for the job.

Employers should focus on involvement and participation of these interns leading to more productive interns, an atmosphere of trust, better solutions and higher standards of service. Interns are more committed and motivated, hence more productive, when actively involved in problem solving. They see their opinions are valued.

Indian hospitality industry's future lies in its young hotel management students and there are various hospitality education providers who have limited knowledge about industry requirements but there are few who provide education that is among the best in the world. But is their education giving them the skills they need to thrive in the workplace once they graduate? The answer appears to be a resounding no. Indian hospitality education providers and policy makers have to plan their intake of students in coordination with the industry. If the imbalance continues the situation of exploitation will not change.

Indian hospitality firms need to learn a lot from internship experiences in UK. Interns need to be viewed as a valuable resource rather than a liability. A firm can always exercise a choice on the kind of individuals it would like to recruit. Long hours, lack of minimum wages, odd jobs at work, lack of decision making powers, lack of independent assignments–are some of the issues which need to be addressed in Indian hospitality firms. The industry and academic awareness needs to be created to have positive experiences for interns. The government and concerned departments or ministries must specify some guidelines on managing these internships. It should be mandatory for firms in hospitality industry to create better work conditions for interns and view them as future stakeholders rather than a short twenty week association.

The recruitment heads at the hospitality firms both in India and UK could mange the programs in three phases–prior to training, during training and after the training. Adequate documentation on expectations from interns, creating brand equity prior to recruitment, clarity on norms for recruitment, transparency on compensation and wages and creating mentors at workplace for these interns would be extremely helpful. During the training, making them go through development programs, creating challenges for them at the workplace can be extremely helpful. Their fresh outlook could be utilized and could be a source of innovation at the workplace. It gives an organization a chance to look at the organization from an outsider's perspective. It gives an organization a chance to assess the potential of an intern as a future employee. It also gives a chance to organization to build brand equity. Post training, an organization could use intern to send communication about the firm. This is a wonderful opportunity for the firm to create wonderful brand ambassadors for themselves. The hospitality industry needs to have policies in place for aspects such as–clarity on internship program, defining internship experience, work content, compensation, support, training and development, future possibilities of alliance. The educational institutions need to have documented procedures for internship, outcomes of internship–expectations from interns and industry, setting up the liaison managers, working closely with the industry professionals. Subsequent studies could be carried out to study whether positive experiences resulted in employment with the same firm and whether negative experiences led the intern to switch over to other firms or industry. Studies could also be carried out to benchmark the internship experiences in hospitality industry.

AUTHOR NOTES

Vinnie Jauhari is M.Sc. (Hons.) in Electronics from Panjab University, Chandigarh and MBA (Gold Medalist) (Marketing). She received her Ph.D. on Corporate Entrepreneurship from IITD and her Post-Doc from the United Nations University, Tokyo in the area of Technology and Society. She has worked at IILM before she joined Institute for International Management and Technology as Associate Professor and Associate Dean. Dr. Jauhari is currently Professor and The Head of the School of Management & Entrepreneurship at IIMT. She is also the founding editor of the Journal of Services Research, has over 45 publications in National and International Journals and has also authored books on Business Strategy and Services Management. Her area of expertise is Corporate Entrepreneurship and Strategic Management.

Kamal Manaktola is MA Public Administration, MHCIMA (UK), Diploma in Hotel Management PGDTD, Certified Master Trainer. He has been trainer in the Hospitality Industry for over 20 years. He holds degrees/diploma in Hotel Management, Public Administration and training & development. He was the Deputy General Manager (HRD) at the Manpower Development Centre at ITDC for over five years. He has worked as a training specialist with European Commission funded South Asia project. He has extensive experience of working in managerial positions at various hotel properties in India and has conducted training programmes for over 1200 hospitality and tourism executives. He has to his credit a number of certificates for completing professional programmes conducted by Cornell University, USA, American Hotel & Motel Association, TMI Denmark, APO-Japan and ISTM-Government of India. Currently he holds the position of Head of School, School of Hospitality and Tourism Management at IIMT Gurgaon.

REFERENCES

Carmen, Sandra (2002), Career Focus: The Hospitality Industry, *What's New*, Philadelphia, Jan/Feb, *35*(3), 21, 2.

Downey, J. E. & De Veare, L. T., (1988), Hospitality Internships: An Industry View, *Cornell Hotel & Restaurant Quarterly*, Nov. *29*(3).

Harris, Kimberly J. & Zhao, Jinlin (2004), Industry Internships: Feedback from participating faculty and industry experience, *International Journal of Contemporary Hospitality Management*, *16*(7), 429.

Herzberg, F., Mausner, B. and Synderman, B. (1959), *The Motivation to Work*, New York, John Wiley.

HR magazine (1997), Internships identify promising employees among college students, *HR Magazine*, April, *42*(4), 102.

Jauhari, V. & Misra K., (2004), *Services Management: An Insight into Hospitality Industry*, Institute for International Management & Technology, Gurgaon.

Jauhari, V., (2001), Employee And Customer Management Processes For Profitability–The Case Of Hewlett–Packard India, *Journal of Services Research*, *1*(1), 149-159.

Le Bruto, Stephen, M. Kenneth, T., (1994), The Educational Value of Captive Hotels, *Cornell Hotel & Restaurant Quarterly*, *35*(4), 72, 8.

Mariano, W., (2002), Several Orlando, Fla. Hospitality Employees Use International Intern Program, *Knight Ridder Tribune Business News*, Washington, April 26, 2002, 1.

Okeiyi, E., Finley, D. and Postel, R. T. (1994), Food and Beverage Management Competencies: Educator, Industry and Student Perspective, *Hospitality and Tourism Educator*, *6*(4), 37-40.

Strock, Clancy (1991), Internship Programs that Work, *Agri Marketing*, October, (29), 9.

Tackett, J. Wolf, F. & Law, D. (2001), Accounting Interns and their employers: Conflicting Perceptions, *Ohio CPA Journal*, Columbus, April-June, *60*(2).

Tanke, M. L. (1988), Course Design for Multicultural Environments, *Cornell Hotel & Restaurant Quarterly*, August 19, *29*(2), 87.

Tas, R. F. (1988), Teaching future managers, *Cornell Hotel and Restaurant Quarterly*, August, *29*(2), 41-43.

Zeithaml, V. A. & Bitner, Mary, J. (2000), *Services Marketing*, Boston, Irwin McGraw Hill.

doi:10.1300/J369v09n02_11

Index

T - #0545 - 101024 - C0 - 216/152/12 - PB - 9781560221456 - Gloss Lamination